V2V/V2I Communications for Improved Road Safety and Efficiency

Other SAE books of interest:

Automotive E/E Reliability
By John Day
(Product Code: T-126)

Automotive Software Engineering
By Joerg Schaeuffele and Thomas Zurawka
(Product Code: R-361)

Vehicle Multiplex Communication
By Christopher A. Lupini
(Product Code: R-340)

V2V/V2I Communications for Improved Road Safety and Efficiency

Edited by Ronald K. Jurgen

SAE International®

Warrendale, Pennsylvania, USA

SAE International®

400 Commonwealth Drive
Warrendale, PA 15096-0001 USA

E-mail: CustomerService@sae.org
Phone: 877-606-7323 (inside USA and Canada)
724-776-4970 (outside USA)
Fax: 724-776-0790

ISBN 978-0-7680-7725-4
Library of Congress Catalog Number 2012939984
SAE Order Number PT-154
DOI 10.4271/PT-154

To purchase bulk quantities, please contact:
SAE Customer Service
E-mail: CustomerService@sae.org
Phone: 877-606-7323 (inside USA and Canada)
724-776-4970 (outside USA)
Fax: 724-776-0790

Visit the SAE International Bookstore at http://books.sae.org

Dedication

This book is dedicated to my friend Richard Keaton.

Table of Contents

About the Editor

Introduction

Introduction

New Driver Accident Avoidance Aids Are on the Way

The annual automobile accident statistics in the United States are sobering: 1.6 million rear-end crashes, 634,000 side crashes that occur at intersections, and 431,000 crashes caused by cars changing lanes or drifting in a lane [1]. These numbers are expected to decrease considerably in the future when new vehicle-to-vehicle (V2V) and vehicle-to infrastructure (V2I) systems become available. Such systems are under development worldwide by major automakers, government, and universities. They show promise for such functions as intersection assist, left-turn assist, do-not-pass warning, advance warning of a vehicle braking ahead, forward-collision warning, and blind-spot/lane-change warning [1].

This book opens with a series of overview news stories and articles from SAE International publications on the progress in this work. This is followed by a series of papers on V2V and V2I dealing with the many technical aspects of design of these systems as well as discussions of such key issues as the need for extreme reliability assurances and traffic congestion overloads on the systems. The following topics outline the key challenges and conclusions cited in some of the articles and papers in this book.

Traffic congestion will be a challenge if, for example, every vehicle in a traffic jam reported in simultaneously, potentially overloading the network. The German auto industry and road ministries are conducting a large-scale test of the technology in 2012 to show that V2V and V2I technologies can operate with "bulletproof" reliability.

The effectiveness of using various configurations of antennas and receivers in studying a four-way blind intersection concluded that more representative regions such as urban areas needed to be tested and examined. Buildings surrounding an intersection can influence the reception in the side streets. Conversely, lack of buildings in certain directions of the intersection can lead to missing reflection surfaces, resulting in considerably less signal power getting into the crossing street. High transmission power can cause radio interference as well.

With V2V technology, vehicles can communicate and exchange information using Global Positioning System capability and technology similar to Wi-Fi. If a V2V-equipped vehicle brakes suddenly, this event can be relayed back to a following V2V vehicle, which can then trigger an alert such as a flashing display or beeping warning to the following driver.

Since future cooperative vehicle safety applications are expected to be mainly communication-based, a significant challenge is in combining and processing enormous amounts of information from the host, surrounding vehicles, and infrastructure in real-time fashion. Dedicated Short Range Communication (DSRC) is a promising protocol of choice for vehicle safety applications.

Features exploiting V2V and V2I communications are still in the early stages of research and development, but growing attention to system-wide infrastructure will lead to the popularity of such features in the future. This will require original equipment manufacturer collaboration on interface and protocol standardization and government-supported road/wireless infrastructure.

References

1. "Stopping Crashes with Smarter Cars," *Consumer Reports*, April 2012, pp. 20–23.

Overviews

On the cusp of connected cars

An auto consortium is set to put the wireless V2X safety network to the test with 120 cars in the Frankfurt region.

by Steven Ashley

A car equipped with V2X wireless communications would automatically alert nearby vehicles if it should become disabled.

Connected-car technology, if done right, would be safe, smart, and affordable. Vehicle-to-vehicle (V2V) and vehicle-to-infrastructure (V2I) communications can alert drivers—and each other—of unseen road hazards and traffic jams. But before cars can be linked in wireless networks, engineers must show that the collective V2X technologies operate with bulletproof reliability. After all, fully verified safety is the only way to earn motorists' trust.

One significant remaining challenge to V2X technology, for example, is traffic congestion. What if every vehicle in a jam reported in simultaneously? Would the flood of signals overload the network? The German auto industry and road ministries are planning to find out during the coming spring in a large-scale test of the technology. Evaluation will take place amid the real road traffic of the some
5 million plus inhabitants of the Frankfurt-Rhine-Main area of the state of Hesse, the country's second largest metropolitan area.

The simTD project (Safe and Intelligent Mobility Field Test Germany) results from collaboration among 18 project partners including major German automakers; suppliers **Bosch** and **Continental**; **German Telekom**; several research institutions and universities; as well as three government ministries.

A 120-vehicle test fleet will begin six or seven months of field trials in the spring, said Christian Weiss, the project coordinator and Manager of Cooperative Systems for the Research and Advanced Development department at **Daimler** AG. Right now several passenger car models—**Audi** A4, **BMW** X1, **Mercedes** C-Class, **Ford** S-max, **Opel** Insignia, and **Volkswagen** Passat—are being fitted to communicate wirelessly with each other and with sensors in road beds and infrastructure via short-range V2X links.

The test zone includes hundreds of ITS roadside stations installed by the Hessian traffic center (VZH) and the Integrated Traffic Management Center Frankfurt (IGLZ) that will enable simTD test fleet vehicles to exchange data with traffic lights, road signs, and traffic control centers.

Next step in auto safety

"We are convinced that car-to-X communication represents an important step on the way to accident-free driving," Weiss noted. European research generally coincides with a **NHTSA** (National Highway Traffic Safety Administration) report that in the U.S. four-fifths of vehicle-on-vehicle accidents involving unimpaired drivers could possibly be prevented if vehicles just talked to one another.

Instrument panel displays under test by the simTD project in Germany will warn drivers of

The simTD field test near Frankfurt, Germany, will evaluate rapid local data transfer among vehicles and roadway infrastructure.

"Foremost for us is the safety benefit," he explained. V2X provides the basis for all kinds of warnings of dangerous road conditions, traffic jams, construction sites and obstacles, and weather dangers. Crucially it can inform drivers early enough to allow them to adapt, to change their behaviors, to avoid hazards. The simTD system is to alert drivers of approaching emergency vehicles, display to drivers the right lane to take for the next turn, or advise on the optimum speed to catch a wave of green lights.

"It's the only sensor that can let you know that there is a hard-braking vehicle right in front of that big truck that's just ahead," Weiss said. Though some radars try to pass under vehicles, they don't have the same potential, he said: "No other sensor can reliably warn you of what's going on just ahead of a truck."

Enhanced traffic management

Traffic efficiency should improve as vehicles transmit information on traffic conditions to a control station, which can then predict and manage traffic developments, Weiss continued. "V2X technology would allow operators to get a current view of the state of the traffic."

Today, magnetic induction loops buried in the roadway or overhead video cameras can count traffic flow at hot spots, but they are expensive and few. Operators, he explained, have to guess what's happening in between sensors.

"V2X provides an accurate view of what's happening on the road network, which allows you to adapt your traffic management strategy to improve capacity utilization, so as to avoid having to build new roads, which is a huge overall challenge," he said. Managers might alter the speeds on variable traffic signs to boost safety or traffic flow.

And then there is the opportunity to piggy-back all kinds of other local services onto the existing network, he noted. Concierge-level mobile services, such as parking space reservations in garages, might soon follow initial installation, for instance.

Short-range wireless

Whatever their name, V2V, V2I, V2X, or Car2X networks are based on heartbeat-like vehicle-status signals that transfer data over short ranges between transponders on vehicles and infrastructure. The simTD's ITS G5 wireless technology, which is tailored to automotive applications, is based on the familiar WLAN standard. The hybrid system meshes the workings of the specially developed wireless vehicle communication standard 802.11p and UMTS mobile phone technology as well as ad hoc networking. This approach was chosen as the most promising because of its potential for favorable economies of scale, he said.

For most applications, the messages are short, but they have to be delivered very rapidly in the tens of microseconds range. The signals only need to travel a maximum of 500 m (1640 ft). For longer distances, the system uses multihopping technology. The wireless message either jumps to a roadside unit, which passes it on to following and oncoming vehicles. These then pass useful messages to others that they meet.

A car fitted with V2X wireless communications will receive early warnings of nearby V2X vehicles that become disabled.

Weiss said that he expected to start the "pre-experiments" to tune the trial's evaluation system as soon as the instrumented fleet expands. **NEC** Laboratories Europe recently delivered the key components of the V2X network software to the simTD partners that enables real-time dissemination of data for traffic safety and traffic efficiency as well as infotainment applications.

A car with V2X can detect icy surface patches using ABS and stability control sensors and then warn nearby vehicles of the hazard.

Real-world tests

Specialists from the **Technical University of Munich** are managing the simTD field test and will evaluate the prodigious amounts of data it should generate. Teamed with researchers from **Wurzburg University**, they are simulating the impact the introduction of the technology would have on traffic if the proportion of cars that were equipped grew enough.

"We have a huge test region, as large as any yet tested, that includes all varieties of infrastructure, all major road types—an airport, a trade fair, high-traffic areas," Weiss said. Analysts will determine how drivers adapt to the technology and establish how successful it is on highway, rural, and urban roads.

"We need to test to [know] what happens if you have a four-lane highway with hundreds of vehicles in traffic transmitting at the same time," said the simTD project coordinator. "The communications must be ensured, even under high load. Scalability is the key," he emphasized, pointing out the costly investment in infrastructure and vehicle technology such a nationwide effort will require. "Plus we don't want to have to update it every minute."

The $92 million (€71 million) project is backed by public and private funding, including support from Germany's **Federal Ministry of Economics and Technology**, the **Federal Ministry of Education and Research**, and the **Federal Ministry of Transport, Building, and Urban Development**.

SAE International feature article reprinted from July 5, 2011 Automotive Engineering International - Online, Volume 1, Number 3

BMW demonstrates left-turn drivers' aid

Depending on the actions of the drivers piloting the oncoming cars, turning left at a busy intersection with no left-turn traffic signal can be a dangerous exercise. Should the approaching traffic include a motorcycle, the potential hazards can easily rise as the small vehicle can be hidden by road features or remain "unseen" by the turning driver because of inherent weaknesses in human perception.

Although they vary according to country and region, traffic statistics indicate that from 30% to 60% of all injury-causing auto accidents and up to one-third of all vehicular fatalities occur at road junctions. Whatever the case, if such a collision does occur, the driver making the turn is almost always held at fault.

Engineers at **BMW Group** Research and Technology in Munich recently demonstrated a left-turn assistant autonomous driver's safety aid to help overcome the difficulties motorists encounter when performing this common maneuver. The system, which the researchers are evaluating in a 5 Series sedan, is intended to remove much of the peril of making turns into complex intersections by addressing both the visibility issues with sensors and the sometimes-tricky decisions regarding whether to proceed with automated braking in the hope of improving safety.

"We've developed an assistance system that helps drivers when they turn to the left by warning them of unseen, oncoming traffic," said Project Manager Arne Purschwitz. "If necessary, it can prevent collisions by automatically applying the brakes."

The novel technology is being pursued as part of the European Union-funded Intersafe 2 initiative, which aims to develop and demonstrate a cooperative intersection safety system that uses sensor fusion data and vehicle-to-vehicle (V2V) wireless communications to reduce accidents at road junctions. Intersafe 2 is a €6.5-million, cost-shared R&D program that is being conducted by a consortium of 11 European car makers, suppliers, and research institutes including BMW, **NEC Europe** Network Laboratories, **Swarco Traffic Systems**, **Volvo Technology**, and **Volkswagen**.

Left-turn oversight

The BMW system activates automatically as soon as the car's sensors determine that the vehicle is entering a left-turn lane and it registers the driver's intention to turn left. Recognition of the lane location is accomplished in two ways, according to company researchers. The GPS sensor in the car's navigation system establishes the car's position within a meter and then matches that point with known intersections on its digital map, while a camera-based image-recognition system reads the road markers that indicate the presence of a left-turn lane.

When the driver engages the car's turn signal, the system triggers three front-mounted laser scanners that map the area forward out to a range up to 325 ft (100 m). The left-turn assistant uses that data to identify oncoming cars, trucks, and motorcycles. If the sensors detect vehicles approaching from the opposite direction but the car continues into the intersection, the system sounds a warning alarm and presents hazard symbols on the instrument panel and head-up displays, while immediately applying automated braking to prevent a collision.

This autonomous intervention deliberately occurs with no prior warning, the researchers said, because an instant response is vital to prevent an accident in these circumstances. Braking is instituted only up to speeds of 6 mph (10 km/h), which leaves drivers free to cross the traffic stream at high speed without electronic controls if they so choose.

As soon as the driver touches the brakes, however, the system halts any automatic braking input, releasing the car so normal driving can continue. To maximize safety the driver can always override the left-turn assistant. If the driver needs, for example, to guide the car to the roadside to clear the way for an emergency vehicle, he or she can do so at any time merely by hitting the gas pedal.

V2V enhances benefit

If the left-turn assistant is augmented with V2V communications, which enables cars to "know" where other vehicles around them are located, greater safety will result, according to company spokespersons. The BMW 5 Series test sedan is fitted with a V2V unit, which boosts the range of its vehicle-recognition function to 820 ft (250 m). It also allows the safety system to detect the presence of unseen vehicles that are equipped with V2V. This capability can be helpful, for instance, when a vehicle

BMW's prototype left-turn assistant, especially if accompanied by V2V wireless communications, could make maneuvers at intersections much safer.

follows a line of cars through a left turn, in which case its sensors would probably be screened by the preceding cars.

BMW engineers also recently conducted another test scenario that highlighted the benefits of this combined function. In this case, the research car approached a BMW R 1200 GS motorcycle fitted with a V2V unit. As before, the data provided by the camera-based image-recognition system and laser scanners enabled the left-turn assistant to register the lane markings, the left-turn arrow, and the distance to the center line and stop lines. When the turning driver engaged the car's turn-signal indicator, the system activated.

"The car and the motorcycle communicated with one another via the car-to-x [V2V] interfaces as the motorcycle approached," explained Udo Rietschel, a development engineer on the project. "The car and motorcycle exchanged information on the type of vehicle, its position and speed, as well as dynamic data such as its steering angle and whether the indicators were activated." The motorcycle's safety system then employed this information to determine that the car driver planned to turn left and move in front of it. An algorithm then calculated the vehicles' trajectories and decided if a collision was likely.

In critical situations, the motorcycle would take measures to warn the car driver by taking steps to raise its visibility—by intensifying the brightness of the cycle's headlights as well as other lights on its sides and mirrors. If its safety system calculates an acute risk of collision, the motorcycle's horn would also sound. Should the car continue entering the intersection, the left-turn assistant would brake the car automatically while activating the appropriate driver alerts and warnings.

BMW's left-turn assistant system, which is still at the developmental stage, was recently demonstrated publicly at a closed-off road junction in Wolfsburg, Germany. If the engineers can prove the viability of the technology, the company could potentially introduce it into production vehicles by 2016.

Steven Ashley

Safer cars talk to each other

Imagine that you are driving at speed behind a big truck. Unknown to you, a car precedes the truck. Suddenly the driver of the lead car hits the brakes hard. Whether you rear-end the truck depends on three things: your speed, the distance to the truck, and how fast you can react to the illumination of its brake lights. One thing that has little to do with your outcome is whether your car has collision-avoidance radar or not. That's because, just like your eye, radar can only sense objects that it can see.

Say, on the other hand, that 10 times a second each of three vehicles is broadcasting a safety status report via short-range Wi-Fi. Suddenly the lead driver slams the brakes. This time the lead car's emergency stop signal and GPS location is transmitted instantly to your car's own digital safety monitor, which rapidly calculates that a collision is imminent. The monitor immediately triggers an unmistakable stop sign—a strip of bright red LEDs on your dashboard. Your chances of missing the truck have improved considerably.

Ford is developing vehicle-to-vehicle (V2V) communications technology that would, for example, be able to warn drivers that a vehicle is blocking the road on a blind turn.

Such advanced safety technology, known as vehicle-to-vehicle (V2V) communications, lies just around the bend, according to Michael Shulman, a technical leader at Ford's Active Safety Research and Advanced Engineering Department, who is heading the company's effort to field-prototype V2V-equipped vehicles within a couple of months.

"V2V can improve the driver's situational awareness through advisories, notifications, and warnings of unseen road hazards," he stated, citing a report issued last fall by the U.S. NHTSA that indicated that each year several million vehicle-on-vehicle crashes, around fourth-fifths of all such accidents involving unimpaired drivers, could possibly be prevented "if vehicles just talked to one another."

Beyond sudden stops, the new technology could alert drivers when two cars are on a collision course at a blind intersection or when a nearby driver changes lanes without looking or loses control. This cooperative wireless link could also warn motorists of less critical issues such as an imminent yellow light or traffic congestion or road conditions.

And not only can V2V help save drivers, it can do so affordably, Shulman said. "The nice thing about V2V communications is that it's relatively low-cost. No exotic technology is needed, only a special flavor of Wi-Fi communications and GPS."

The other big piece of the puzzle, the necessary digital control smarts, have already been developed for collision avoidance and automatic cruise control (ACC), radar-based systems that can cost thousands of dollars to install, he continued. "This means that all cars, not just luxury ones, can have this capability; we want to sell this technology on the Fiestas as well as the Lincolns."

Beyond safety

Although enhancing safety—accident avoidance—is Ford's priority, V2V can also bring other significant customer benefits in a variety of applications related to eco-mobility, infotainment, and driver convenience, Shulman observed. "We want to be the leader in network-connected vehicles and in bringing these new benefits to the customer." He noted that Ford management has doubled its investment in intelligent, connected vehicles and has assembled a task force of 20 personnel, including planners, engineers, and scientists, to accelerate development of the new comm link. "We plan on becoming the first automaker to build prototype vehicles for demonstrations across the U.S. in the spring," he said.

Shulman added that his team is busy readying eight prototype Taurus sedans with V2V communications systems that are targeted for delivery to the U.S. DOT early this summer. The cars will be part of a fleet of 64 equipped vehicles that are being supplied by a variety of makers for a government/industry pilot program to evaluate and road-test V2V technology. Also included in the pilot effort will be V2V's cousin, vehicle-to-infrastructure (V2I) systems—vehicle communication with road fixtures such as traffic signals and toll booths. As part of the demonstration program, the DOT will sponsor development of inexpensive aftermarket V2V devices that can be retrofitted to existing vehicles.

The DOT is mounting the R&D effort because its administrators believe that the technology provides one of the most effective pathways to improved road safety. "This technology is an opportunity to help create a future where millions of vehicles communicate with each other by sharing anonymous real-time information about traffic speeds and conditions," said Peter Appel, head of the DOT's Research and Innovative Technology Administration. "This new world of wireless communication will make transportation safer, provide better and faster exchange of information for vastly improved daily and long-distance travel, and even reduce environmental pollution."

Also taking part are state and local highway authorities, who are looking for ways to obtain good, real-time traffic information from these distributed networks. This data would truly help them manage congestion and improve mobility on their roads, Shulman said. "If drivers could report their recent travel history in a private way, the authorities could react to problems much faster and the network could provide timely alerts to motorists."

Crossroads for V2V

This year seems to mark a crossroads for V2V and V2I technology as recent advancement is the culmination of years of cooperative, precompetitive work by industry and government to prepare the promising safety technology for introduction. The current state of progress was arrived at through eight years of joint efforts by car companies (Daimler, Ford, General Motors, Honda, Toyota), government agencies (DOT and NHTSA), and professional organizations (SAE International and IEEE). The initial industry-wide cooperation came naturally because of the inherent cooperative nature of this technology, especially given that the benefits of V2V and V2I will only start to manifest themselves when a reasonable fraction of the vehicles on the road are suitably equipped.

The joint V2V effort started in 2002 when the FCC allocated radio spectrum for dedicated short-range communications (DSRC) to enable cars to talk to other cars and the road infrastructure (a 75-MHz spread on a 5.9-GHz carrier wave), Shulman recalled. "The concept was to communicate within 300 meters, and unlike radar, you could take advantage of the fact that V2V doesn't need direct-line of sight," he said. "To convince ourselves of its feasibility, the five companies did tests like sending 1000 packets at speed in bad weather to see how many were received," he explained. The rigorous evaluations showed that the link "was pretty robust, which persuaded us that V2V will really work and is worth implementing."

The industry, he continued, agreed to build V2V using a version of Wi-Fi that was described in the IEEE 802.11p standard, a carrier-sense multiplex-access protocol. Other relevant standards are an interim IEEE 1609 protocol and SAE J2735, which "has to do with what should be in the basic message set." Working with DOT experts, the coalition decided that a message would include such things as a vehicle's position, speed, brake status, path prediction, path history, vehicle mass, and bumper height.

"This message is sent out 10 times a second by every equipped vehicle, which enables your system to place and plot out surrounding vehicles in space and receive messages about potential hazards," Shulman explained. Along with NHTSA specialists, company engineers identified traffic scenarios in which V2V could improve safety: forward collision warning, emergency electronic braking, blind spot warning, lane change warning, do not pass, and control-loss warning. He stressed that "V2V and V2I allow safer operations at intersections, a place where it's really tough to improve safety."

A secure and private network

"Of course, we don't want anybody tracking drivers or any law enforcement involvement," he said, "so everyone's ID only lasts five minutes and then it changes." This scheme retains enough short-term linkability to deliver services but maintain anonymity. The designers also ensured that the network could maintain security from attack by malicious users or hackers by installing PKI (public key infrastructure) cryptographic systems.

Shulman said that the public-private partnership has a clear road map to finish up all remaining research and standards and then move on to institute regulations ("a rule-making process"). "We still need a standard to be set on congestion management, that is, what happens when you're in traffic and there are 100 cars within 300 meters. If everybody transmits 10 times a second you fill up the channel," he explained. The system will probably reduce power or message frequency in that case.

"V2V is a very interesting application," Shulman concluded. "Current developments show that the industry is transitioning from cooperation to competition. Everyone must cooperate to make the system work to improve safety, but here at Ford we're also trying to differentiate ourselves by focusing as well on the human-machine interface as well as a range of unique and exciting convenience apps for the customer."

Steven Ashley

SAE International feature article reprinted from March 16, 2010 Automotive Engineering International - Online, Volume 1, Number 3

V2V, GPS integration could improve safety

The current drive to employ active safety systems to prevent crashes is based on radar, cameras, and other sensors, but there is a powerful weapon looming in the wings. Vehicle-to-vehicle communications and global positioning satellites are being examined to play a role in the push to reduce traffic accidents.

Automakers and Tier 1 suppliers are rapidly deploying radar systems and cameras, combining their input for active safety systems that can, among other things, apply brakes automatically when a collision is imminent. The moves are part of a global effort to reduce fatalities and damage.

Regulators such as the **NHTSA** (U.S. National Highway Traffic Safety Administration) are putting more emphasis on preventing crashes, feeling there is more benefit to be gained by avoiding crashes than in continuing improvements in passive safety systems that respond after the fact. Automakers are planning to add this input to active safety systems, treating it much like other sensor data.

"We think of vehicle-to-vehicle [V2V] communication as another type of sensor that will ultimately show up, though probably not until a number of vendors deploy it so you have more vehicles on the road communicating," said John Capp, **General Motors** Co. Director of Global Safety Technology.

Concerns about the expense of putting intervehicle transceivers into vehicles that will not have anyone to talk to is a major question surrounding the rollout of these products. The **U.S. Department of Transportation** is making V2V a key element of its Intelligent Transportation System plans.

One facet of the plan is intended to speed up the rollout so that a critical mass of vehicles will begin improving safety by warning other vehicles that they are nearby. DOT thinks consumers will buy aftermarket products that could be used in older vehicles.

"Aftermarket devices could emit 'Here I am' messages that are then used by IntelliDrive V2V and vehicle-to-infrastructure applications," said Shelley Row, Director of the U.S. DOT's Intelligent Transportation Systems Joint Program Office. "This would make it possible for older vehicles to be a part of a nationwide network of connected vehicles, not just newer vehicles that have factory-installed equipment."

That could bring substantial improvements in safety. However, Row noted that vehicles with aftermarket devices would not receive the same safety benefits as vehicles with factory-installed equipment, since many safety applications require interaction with vehicle systems such as braking.

Many researchers are also looking for ways to leverage the rapidly expanding telematics industry to speed up the technology's growth. Telematics transceivers could be altered to include V2V compatibility. However, some market watchers feel that government incentives beyond DOT efforts may be needed.

"Telematics service providers will only participate if it's easy, it comes at no cost, or brings some benefit to them," said Thilo Koslowski, Automotive Vice President at **Gartner**. "The government may give them some sort of tax break for participating. Carmakers are still trying to get people to buy into telematics, so it may also take some tax breaks to get them involved."

While developers create plans for V2V, other projects are planning to leverage another wireless input: GPS data.

"Sensor fusion between radar and cameras can also include GPS," said Martin Duncan, Innovative Systems Manager at **STMicroelectronics**. "You can get a lot of nice information from GPS. It gives you another data point for speed monitoring, and cameras can cross check with GPS systems."

GM's Capp noted that GPS can also be used to augment V2V input that is used by safety systems. When vehicles know where they are in relationship to roads, obstacles, and other vehicles, there is a good possibility that collisions can be dramatically reduced.

"V2V is a 360-degree sensor. Combining it with GPS so you know where the vehicles are puts you on the way towards a crashless vehicle," he said.

Terry Costlow

Vehicle-to-vehicle communications may become an input for active safety systems. (U.S. DOT)

Debating IntelliDrive's future

Though the government's decision on whether to actually deploy IntelliDrive won't be made until 2013, there's an ongoing debate over how quickly the communications technology could begin making an impact. Building a critical mass of vehicles will take time, and maintaining the infrastructure could be costly.

The U.S. Department of Transportation is still developing the wireless networking technology and determining how it can improve safety, reduce congestion, and conserve fuel. In 2013, government officials will decide if and how IntelliDrive will be deployed. Many observers feel that, given all the time and money that has already been invested, there is a high likelihood that the technology will be deployed.

If the communications network does move forward, a time frame for its impact is more nebulous than its rollout. Both the vehicle-to-vehicle (V2V) and vehicle-to-infrastructure (V2I) require a fair number of vehicles to provide significant benefits.

"The first people to buy cars with vehicle-to-vehicle communications will be pretty lonely," said Thilo Koslowski, Automotive Vice President for Gartner Inc. "IntelliDrive will need a critical mass to make it useful, we need to get 10-20% of the vehicles on the road so data will be useful."

That critical mass might be achieved more quickly if dedicated short range communications (DSRC) transceivers are piggybacked into telematics hardware, which is expected to have reached significant volume by the latter half of the decade when IntelliDrive may start moving into the field.

Many telematics system suppliers are open to the idea of adding transceivers into their modules. "We're agnostic. If there's a national network for DSRC and it makes sense to put that in the telematics box, why not?" said Erik Goldman, President of Hughes Telematics.

However, there is still a big question as to who will pay for these transceivers, which will communicate on a dedicated 5.9-GHz channel. Neither automakers nor telematics suppliers are likely to voluntarily include the communications chips, and consumers aren't likely to pay for communication devices that may not provide benefits for a few years.

"Some sort of stimulus will have to come from the government," Koslowski said. "Consumers won't be willing to pay for it until they see a real, tangible benefit."

There is concern that a DSRC infrastructure will be expensive to install and keep up. Two IntelliDrive goals, reducing traffic congestion and cutting fuel consumption and emissions, require V2I links. Roadside towers will gather data on traffic movement and then compile it to help drivers find travel alternatives or delay journeys until traffic jams abate.

"Nationally, it will be a vast network that needs to be maintained," Goldman said. "Communication technology changes rapidly, so the ongoing expense will be significant."

Some observers question whether DSRC will be needed for roadside monitoring. Telematics system transceivers and cell phones can provide data on vehicle location, which can be analyzed to determine traffic speed and congestion levels.

If telematics expands as quickly as most analysts and marketers predict, a solid percentage of vehicles on roadways will have transceivers by the latter half of the decade. "Getting a critical mass of vehicles will be easy," said John Horn, T-Mobile's National Director of Machine to Machine Communications. "The infrastructure will be able to gather information on the movement of modules."

However, he noted that there will still be a need for many infrastructure receiving stations as well as for powerful servers that will make sense of the data that are gathered. These servers will have to understand what's happening on roadways and send useful, concise messages to drivers so they can alter their plans. That will require sophisticated software.

"Writing the algorithms that analyze all this data will be where the secret sauce comes in," Horn said. "Tools need to be sophisticated enough to know whether traffic is stopped for a red light or whether a lane is blocked."

Terry Costlow

V2V and V2I Technical Papers

2012-01-0741
Published 04/16/2012

doi:10.4271/2012-01-0741
saepcelec.saejournals.org

An Autonomous and Car-Following System via DSRC Communication

Chan Wei Hsu, Ming Kuan KO, Min Huai shih and Shih Chieh Huang
Automotive Research Testing Center

ABSTRACT

Inertia navigation system is capable to backup GPS unavailability. An autonomous system integrated different sensors to offer robots or remotely operated systems a reliable navigation means. This paper addresses an autonomous design method to accomplish continuously position positioning based on GPS positioning, gear-box speed, odometry, IMU and DSRC in deck reckoning. Besides, the availability of autonomous function is verified and applied using vehicles in car following or cooperative driving. This paper also presents an inter-vehicle geocasting format to accomplish multihop and car-following in cooperative driving and data exchanged based on GPS/IMU positioning and DSRC. The car following provides warning or situation awareness to drivers based on broadcast messaging the motions of neighboring vehicles obtained by one-way communication from other vehicles. The proposed system has the advantage of omnidirectional transmitting/receiving functions of DSRC module that provides 360-degree coverage in own surveillance region. The proposed system is carried out with theoretical application and hardware integration, and furthermore the result shows navigation ability and intelligent approach applicability.

CITATION: Hsu, C., KO, M., shih, M. and Huang, S., "An Autonomous and Car-Following System via DSRC Communication," *SAE Int. J. Passeng. Cars - Electron. Electr. Syst.* 5(1):2012, doi:10.4271/2012-01-0741.

INTRODUCTION

Every year in Taiwan, about two thousands deaths within 24 hours in traffic accidents, there are about 2539 deaths per hundred thousands of people and the statistical number is very serious in the world [1]. While many different factors contribute to vehicle incidents or accidents, such as rainy day or blind spot area, driver behavior is considered as the main cause of more than 95 percent. Traffic safety, in terms of infrastructure or injuries, has been discussed and improved by government's policy. However, the numbers of deaths or injuries have remained relatively flat due to the increasing number of vehicles or fatigued driving with low attention. In the recent years, more and more people like to have a team travel in the weekend. People like to take a portable navigation device with them. It provides high accuracy position, any weather condition and has the advantage in faster positioning. Although it is easier to know own location mapped onto GIS, groups of team trip cannot be aware of others position.

The inertial navigation system (INS) can overcome this shortcoming by inertia sensors. The acceleration and spatial information can be obtained from accelerometers and gyroscopes of any moving platform. An INS is an all-weather autonomous navigation system that can provide continuous position, velocity and attitude information in real-time operation [2]. The main defect of the INS is that its mean-square navigation error increases with time and needs frequent calibration with reference signals. INS error accumulates due to inertial sensor's performance with time that long period performance of INS becomes less accurate. A vehicular unit (VU) could sense vehicle speed and heading by calculating odometry. Although lower accuracy inertial sensors might cause the integration error with time in speed and vehicle spatial motion, a VU could provide continuous speed and heading with movement through controller area network (CAN). As a result, an idea of GPS/INS and VU integrated system tries to adopt commercial GPS and inertial sensors to construct a higher reliable and better accurate navigation in lower cost platform. In order to limit INS navigation errors, the INS position information could be updated in accordance with GPS, and vehicular data information is update and enforced by VU data acquisition and computation.

In the context of ITS, wireless communication plays a fundamental role in the recent two decades. In system concept, a vehicle can broadcast its driving parameters to others over Wi-Fi technologies, like dedicated short range communication (DSRC) [3]. The choice of ad hoc network, contrary to cellular network, is more rigorous and justified by the fact that the network is organized without an infrastructure which avoids data blocking or unavailability of the network as in 2.0 to 3.5G mobile communication. To meet a higher vehicle safety, DSRC which has a wireless communication protocol in the 5.9 GHz frequency band plays an important role of vehicular system. Indeed, by communicating information in remote surveillance on possible emergencies, dangerous events can be avoided. Thus, exchanged data can be used to improve the safety and become aware of neighboring vehicles location, including speed, location and heading. In addition, IEEE has taken up the standardization of DSRC by creating IEEE 802.11p [4]. In the proposed system, 802.11p protocol had been porting into embedded system for DSRC data link layer.

This paper adopted an embedded system to construct an independent navigation platform using data fusion integration for driving navigation. Combining with the GPS/IMU integration, the vehicular signals may play a potential auxiliary support to derive another package of position and moving information to enhance the autonomous capability. The hardware has been implemented on microcontrollers and carried out verification tests. The DSRC application and collision design has been presented in ARTC. Besides, the packet and data geocasting method are also designed in the proposed system. The following content will be focused on autonomous design, car-following and its verification.

SYSTEM ARCHITECTURE

The system technology is designed with an integration of INS module and data transmitting through DSRC module. The INS module could provide a good positioning solution, and then the positioning information may display on the screen to monitor other neighboring vehicles in remote operation. The vehicular data of vehicle will be routed by CAN module and broadcast to neighboring vehicles by UDP protocol onto the internet via DSRC communication. The concept of proposed system architecture is shown in Figure 1. The test information is debugged and showed in the screen of laptop using well defined format, and the total lengths which follow CAN 2.0A is about 8 bytes with its id in different devices.

A general vehicular communication which depends on its coverage area can be classified into four categories: inter-vehicle, outer-vehicle, vehicle-to-infrastructure (V2I) and vehicle-to-vehicle (V2V). A common solution, Bluetooth, its coverage is too low to do vehicular communication as a precaution. In outer-communication, mobile communication has presented its wide coverage, high reliability of data transmission in surveillance applications. However, mobile communication has a drawback in time delay about 1.0 sec in TCP mode or 0.8 sec in UDP mode [5]. To meet a high converge, data rate and low time latency. DSRC is a good choice, and its theoretically provides up to a 1 km range and allows communications between vehicles moving up to 160 km/h [6]. It also has low latency about 50 millisecond and 8 priority levels. In DSRC software, the network protocol is based on IEEE 802.11p standard under open system interconnection (OSI) model. This layer is ported from revising 802.11a, and other layers are followed UDP/IP mode.

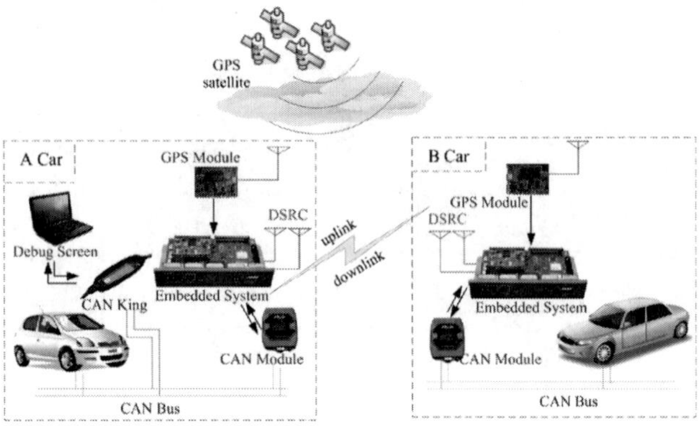

Figure 1. System architecture.

A. THE PROPOSED SYSTEM ARCHITECTURE

A data fusion of car-following system platform is built of the basic three parts, including INS module, DSRC communication unit and vehicle unit. A X86 processor is the mainly adopted controller, where the embedded kernel to access data input and output is programmed. To fulfill the proposed anti-collision application, INS data as well as vehicular data are processed in specific logic, digital formats and sent through DSRC module in controlled intervals. The data packet is formed up in X86 processor from the peripheral sensors as well as INS. Each data packet is collected and sent to DSRC module within each cycle of data surveillance. Meanwhile, the processor is activated by embedded Linux system. The USB of embedded system interface is simulated as a serial port that is used as COM1 port for INS module to capture positioning data, and two mini-PCI slots for DSRC modules to transmit data or receive data from other vehicles.

An INS module, autonomous function, is built of the basic four parts of processing core; Inertial Measurement Unit (IMU) sensors with accelerometer and gyroscope; GPS Receiver and vehicular information unit [7]. The system architecture of INS core is shown in Figure 2. The dsPIC30F6014 [dsPIC30F6014 Datasheet, Microchip Products Inc., web: http://www.microchip.com/] is chosen as the core controller to handle real time message. Time slots are used to process and measure the inertial sensors data and GPS data through Universal Asynchronous Receiver and Transmitter (UART) port.

Table 1. Requirements and specifications of INS Core.

Criterion	Required	Desirable	dsPIC30F6014
On chip nonvolatile memory	Enough to eliminate need	for data storage and use to buffer the data	144kB Flash, 8kB RAM
ADC	6 channels, at least 12-bit resolution, with 100SPS	Simultaneous sampling, ADC buffer memory	16 channels, 12-bit resolution, simultaneous S/H, up to 500k SPS
I/O pins	Enough I/O pins to support transmission	I/O pins available for more buttons	Over 48 I/O pins
Timers	At least one 16-bit timer	More timers	5*16-bit timers available
Controller Area Network interface	Data Acquisition from CAN	Vehicular data available	2 channel CAN, up to 500 kbits

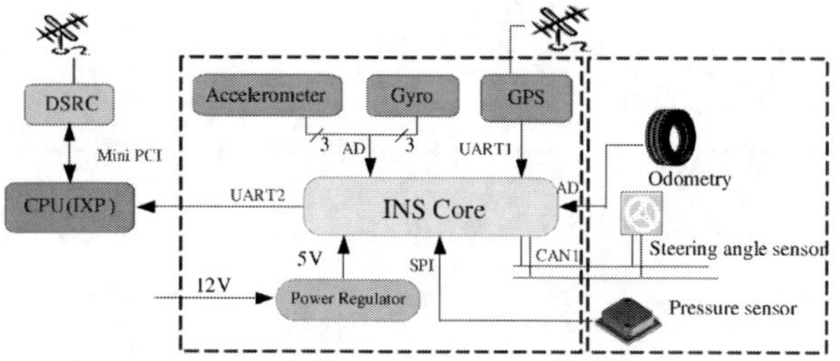

Figure 2. Block diagram of INS.

B. AUTONOMOUS CORE (INS)

The dsPIC30F6014 is Microchip product for signal processing. This chip are designed to perform as supervisor core, where GPS message and inertial analog signals are on-lined captured, sampled and processed, back to the DSRC application; while the vehicular data is determined as the calibrated information. This INS core has the communication interface to a personal computer/IXP to downlink messages and broadcast to adjacent vehicles in real time. As the needs for the system, the specification and requirement of dsPIC30F6014 are listed in the following Table 1.

An IMU is a closed system that is used to detect altitude, location, and motion. It normally uses a combination of accelerometers and gyroscopes to track the vehicle motion in attitude and location. The IMU is capable to transmit inertia data from core to IXP, where the interface is full duplex UART2. In order to output vehicle spatial states, it is accomplished by integrating an output set of sensors, such as gyros and accelerometers. Gyros measure angular rate with reference to inertial space, and accelerometers measure linear acceleration with respect to vehicle's frame. The IMU utilizes a tri-axis accelerometer and three one-axis gyroscopes as inertial measurement components. The accelerometer is measured for X-Y-Z axis; while the gyros are assigned to X-Y-Z axis correspondingly. The IMU plays a full inertial function for vehicle in real time.

The ADXRS614 operates on the principle of a resonator gyroscope. The output signal of ADXRS614 is a voltage proportional to angular rate about the axis normal to the top surface of the package. With the increase of the rotation rate, the output voltage leaves the neutral point [ADXRS614 Datasheet, Analog Devices Products Inc., UK, web: http://www.analog.com/]. An external capacitor is used to set the bandwidth. Use external capacitors in combination with on-chip resistors to create two low-pass filters to limit the bandwidth of the ADXRS614's rate response. ADXL330 is made by the principle of resonant accelerometers. It can measure both dynamic acceleration (e.g., vibration) and static acceleration (e.g., gravity). The outputs are analog voltages proportional to acceleration [ADXL330 Datasheet, Analog Devices Products Inc., UK, Rev. A., 2003. Available on web: http://www.analog.com/]. This sensor is capable of measuring both positive and negative accelerations to at least ±3 g. Because the signal measure from the accelerometer is analog, it may be disturbed by external noise. According to specifications, the external capacitor can be chosen to determine the bandwidth of the accelerometer, e.g., 0.47μF capacitor for 20 Hz is used in this paper.

C. VEHICLE UNIT FROM ON-BOARD DIAGNOSTIC CONNECTOR

The amount of electronic devices in vehicles is diagnosed by CAN bus in recent years [8]. In system platform, data is transmitted or received by CAN bus. CAN is a serial, asynchronous, multi-master communication protocol for connecting electronic control modules, sensors and actuators in automotive and industrial applications. The CAN-based system is based on the broadcast communication mechanism which is achieved by using a message oriented transmission protocol. The bit rate of CAN bus is up to 1 Mbps and is possibly operated at network lengths below 40 meter. In this study, the data rate is 500 kbps and its sampling point is held in 75%.

The CAN bus is designed and built in self-defined protocol, but the data acquisition from OBD should refer to motor standard [9]. To fulfill the proposed application, steering angle data as well as inertial data are processed in specific logic, digital formats and sent through CAN bus in controlled intervals. Each data packet is less than 8 bytes using standard ID (11 bit); moreover, the refresh time of packets is about 20ms. The CAN transceiver is the interface between the CAN protocol controller and the physical

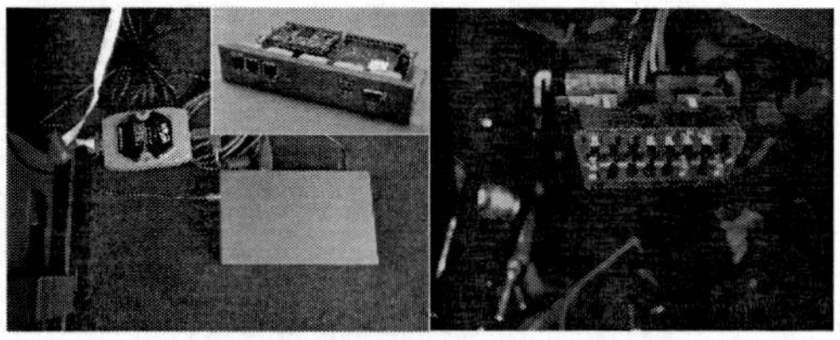

Figure 3. System hardware and OBD II connector.

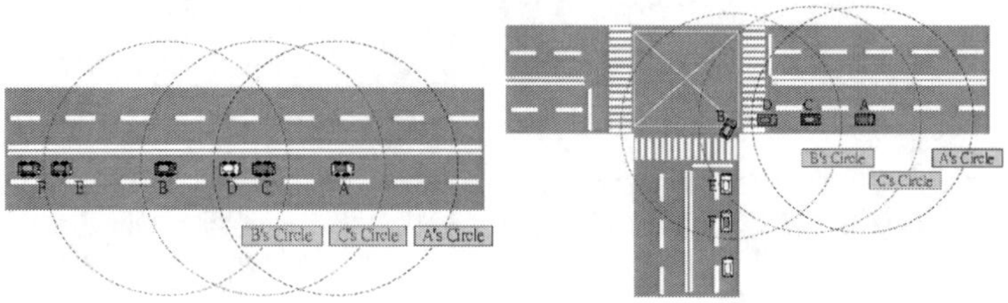

Figure 4. Inter-Vehicle Geocasting in Straight Roadway and Intersection.

transmission line and it is one of the key factors influencing the capability of network system. It is fully compatible with the "ISO 11898".

Each unit is implemented by CAN circuit board, and INS is embedded CAN controller and it adopts NXP TJA 1040 as the transceiver. The CAN transceiver is the interface between the CAN controller and the physical transmission line and it is one of the key factors influencing the capability of network system. The OBD II connector usually locates near brake/throttle under steering, and the connector is D-type and 16 pins with CAN interface. In Figure 3, the left part is system hardware and right is the OBD connector which is used to capture vehicular information.

PRINCIPLE OF SYSTEM ALGORITHM

A. INTER-VEHICLE GEOCASTING IN CAR-FOLLOWING

DSRC has the advantage of omnidirectional transmitting/receiving functions. However, it needs to depend on broadcasting messages. Geocast, i.e. the transmission of a message to some or all nodes within a geographical area, allows promising new services or application [10]. In inter-vehicle geocasting, each vehicle can broadcast own vehicular message and receive messages from neighboring vehicles [11]. In real application, each DSRC module has about 300 m of transmitting ability in this paper. This is active surveillance area, and there is another surveillance area which is about 500 m. Moreover, the outer area is relied on inter-vehicle geocasting. To meet a good geocasting, there are two key points to concern: relative distance and heading difference between prior vehicles and following ones. CCW is sent in the form of sentences; each starts with a dollar sign "#" and terminates with a carriage return <CR> and line feed <LF>. There are 9 parameters setting for CCW including group id, source node, repeater node, UTC time, latitude, longitude, height, heading, and vehicle speed. If the message is original one, the repeater node will be null string. The parameters are arranged as the sentence as follow: "#ARTC,E,, 082714,24.059958,120.383784,8.6,310.62,63.1".

In Fig. 4, left figure shows straight driving in highway or expressway and the right one is intersection case. Taking left Fig. 4 as the example, B node transmits its message periodically and receives other messages from neighboring vehicles. In B's area, it only receives messages from C, D and E. At this moment, it will determine which vehicle is located in the edge of its transmitting area. In the logical decision, B will repeat C and E message in its region if the communication time is small than 2 seconds. The communication time is calculated using Eq.(1) which is from relative position and its projection in relative coordinate. From relative position (x and y), speed (V_X and V_y) and transmitting range (R), the communication time (C_t) is obtained. For C node, it can receive A, B D and E messages from B. The previous segment is only available in straight or low curve roadway, but it cannot communicate with turned vehicle, such as right Fig. 4. In order to avoid this kind of case, the turning vehicle will broadcast vehicles message which have large heading difference relatively. Owing to non-synchronous GPS time, the message parameters include time stamp. The time stamp also uses to check the time difference and update the message by checking the effective messages.

$$C_t = (-(V_x \cdot x + V_y \cdot y) + \sqrt{(V_x^2 + V_y^2)R^2 - (V_x \cdot y + V_y \cdot x)^2}) / (V_x^2 + V_y^2) \qquad (1)$$

The relative position is transformed from WGS-84 to ECEF and ECEF to NED frame using Eq.(2)-(3). The altitude (h) is given by GPS receiver and the other parameters are eccentric (e) and semi-major axis (a). Eq.(2) is result from the shape of the Earth which is an ellipsoid, not a true sphere. The following procedure is to take ownership as center and calculate relative position using Eq.(3). In Eq.(2)-(3), N is the radius of curvature in prime vertical, and two vehicle position are located in (Λ_0, λ_0) and(Λ_1, λ_1).

$$\begin{bmatrix} x^E \\ y^E \\ z^E \end{bmatrix} = \begin{bmatrix} (N+h)\cos\Lambda\cos\lambda \\ (N+h)\cos\Lambda\sin\lambda \\ [N(1-e^2)+h]\sin\Lambda \end{bmatrix} \quad N = \frac{a}{\sqrt{1-e^2\sin^2\Lambda}} \tag{2}$$

$$\begin{bmatrix} x^N \\ y^E \\ z^D \end{bmatrix} = \begin{bmatrix} -C(\lambda_0)\cdot S(\Lambda_0) & -S(\lambda_0)\cdot S(\Lambda_0) & C(\Lambda_0) \\ -\sin(\lambda_0) & \sin(\lambda_0) & 0 \\ -C(\lambda_0)C(\Lambda_0) & -S(\lambda_0)C(\Lambda_0) & -S(\Lambda_0) \end{bmatrix} \times \begin{bmatrix} x_1^E - x_0^E \\ y_1^E - y_0^E \\ z_1^E - z_0^E \end{bmatrix} \tag{3}$$

B. ATTITUDE CALCULATION - QUATERNION METHOD

Fig. 5 describes how to achieve inertial navigation via measurement and frame transform. The navigation algorithm contains several steps to compute vehicle attitude, earth rate, transport rate and Coriolis. The procedure to integrate acceleration, angular rates and calculate vehicle states in hardware is operated with software which is built in the microprocessor following the theoretical formulation below.

Euler angles are the values which present the attitude of the vehicle. The attitudes of the vehicle mean the angles between body axes with navigation axes. There are three Euler angles φ (roll), θ (pitch) and ψ (yaw) used to show the relative angles along x, y and z axis. The Direct Cosine Metrics (DCM) is used to transfer information from one coordinate system into another coordinate system. It is carried out as a sequence of three successive rotations about different axes. The DCM mentioned earlier is limited since the solution become indeterminate when θ is approximately 90 degree. The quaternion is the method to overcome this problem. The concept of the quaternion is based on the idea that a transformation from one coordinate frame to another can be effected by a single rotation angle δ and an orientation unit λ defined with respect to the reference frame in left part of Fig. 6 and right part is the definition of Euler angle.

In quaternion transformation, the orientation is written as a vector which contains four elements with the magnitude of the rotation. The preceding about body-to-navigation DCM can be expressed through quaternion elements as Eq.(4). The q_i (i=0, 1-3) are calculated from kinematic equation in Eq.(5), and it can be substituted with the quaternion elements which is shown in Eq.(6). From Eq.(6), vehicle attitude can be integrated and updated to get Euler angles.

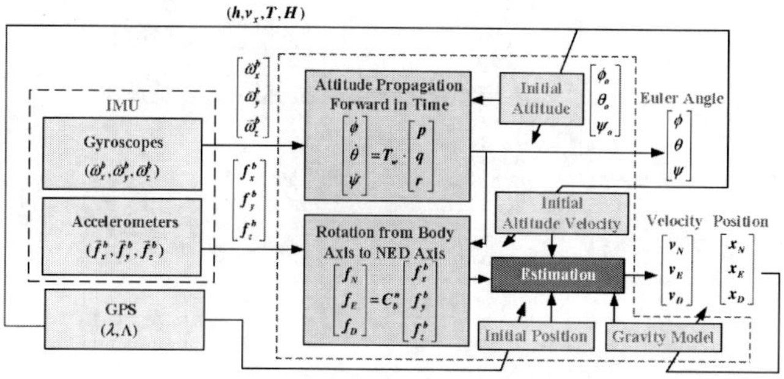

Figure 5. The block diagram of algorithm in navigation.

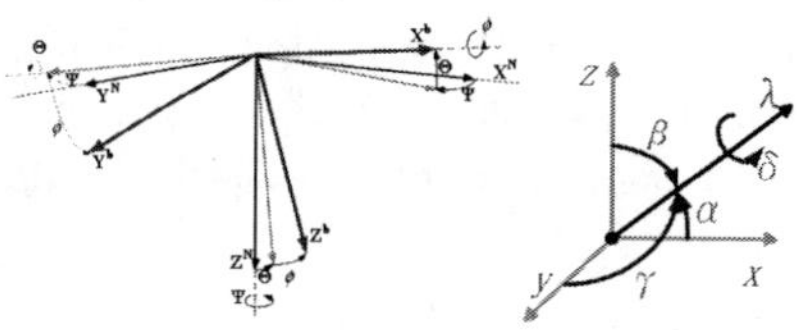

Figure 6. Definition of Euler angles and Quaternion coordinate.

$$C_b^n = \begin{bmatrix} q_0^2+q_1^2+q_2^2+q_3^2 & 2(q_1q_2-q_0q_3) & 2(q_1q_3-q_0q_2) \\ 2(q_1q_2+q_0q_3) & q_0^2-q_1^2+q_2^2-q_3^2 & 2(q_2q_3-q_0q_1) \\ 2(q_1q_3-q_0q_2) & 2(q_2q_3-q_0q_1) & q_0^2-q_1^2-q_2^2+q_3^2 \end{bmatrix} \tag{4}$$

$$\begin{bmatrix} \dot\phi \\ \dot\theta \\ \dot\varphi \end{bmatrix} = \begin{bmatrix} 1 & S\phi T\theta & C\phi T\theta \\ 0 & C\phi & -S\phi \\ 0 & S\phi / C\theta & C\phi / C\theta \end{bmatrix} \begin{bmatrix} p \\ q \\ r \end{bmatrix} \tag{5}$$

$$\begin{bmatrix} \dot q_1 \\ \dot q_2 \\ \dot q_3 \\ \dot q_0 \end{bmatrix} = \frac{1}{2} \begin{bmatrix} 0 & r & -q & p \\ -r & 0 & p & q \\ q & -p & 0 & r \\ -p & -q & -r & 0 \end{bmatrix} \begin{bmatrix} q_1 \\ q_2 \\ q_3 \\ q_0 \end{bmatrix} \tag{6}$$

The variations of velocities are integrated from the accelerations in the local geodetic frame. However, the measurements derived from sensors are in body frame. Therefore, the DCM mentioned earlier is used to transform the measurements from body frame into the local geodetic frame. The transformation is shown in Eq. (7). Owing to the Earth's rotation ($2\omega^n_{i/e}\times v^n$) and gravitation ($g^n$), the effect of the Coriolis force ($\omega^n_{e/n}\times v^n$) and gravity need to be corrected in the middle term of Eq.(7). The position of vehicle is always described with longitude, latitude, and altitude (Λ,λ,h) in local geodetic frame. The navigation systems using on earth surface are mechanized or implemented such that the local geodetic frame is maintained while the vehicle is moving. The ellipsoidal model of the Earth is used to

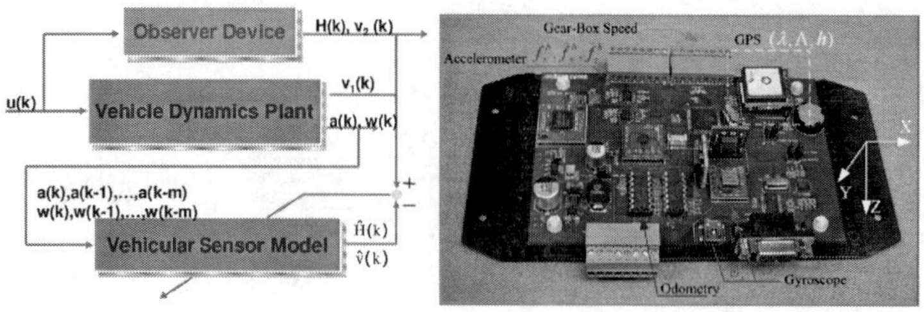

Figure 7. Inertial sensors calibration method and its hardware.

orientate the navigation frame with the variation position of the vehicle. The equations of the ellipsoidal are R_{meridian} and R_{normal}. Motion over the surface of the Earth is along the arc of the ellipsoidal surface. The changing rate of latitude and longitude are expressed in terms of R_{meridian} and R_{normal} in Eq.(8). The variation of altitude is equal to negative down velocity.

$$\dot{v}^n = f^n - (w^n_{e/n} + 2w^n_{i/e}) \times v^n + g^n \tag{7}$$

$$\begin{bmatrix} \dot{\Lambda} \\ \dot{\lambda} \\ \dot{h} \end{bmatrix} = \begin{bmatrix} \frac{1}{R_{meridian}+h} & 0 & 0 \\ 0 & \frac{1}{(R_{normal}+h)C\Lambda} & 0 \\ 0 & 0 & -1 \end{bmatrix} \begin{bmatrix} v_{north} \\ v_{east} \\ v_{down} \end{bmatrix} \tag{8}$$

C. SENSOR CORRECTION - INERTIAL SENSORS CALIBRATION METHOD

The vehicle sensing components consist of accelerometer and gyros, and furthermore they are used to perform the attitude computations in strapdown. These errors are associated with gyros and accelerometer that typically include static biases, drifts, scale factor and random noise. To solve this kind of shortcoming, inertial sensors calibration method is applied in this paper, as shown in Fig. 7. The vehicle plant is demonstrated vehicle, and the output is captured from IMU. In calibrated operation, this paper corrects gyro and accelerometer parameters. Taking gyro procedure as example, the observer device used GPS to get vehicle heading. Refer to Eq.(9)-(10), system controller integrated angle rate and compared with heading. This method uses second-order minimal energy and gradient method to get error variation in Eq.(11). The relation can be derived to discrete form in Eq.(12).

$$\hat{H}[k] = \int (\hat{p}_n w^n + \hat{p}_{n-1} w^{n-1} + \ldots + \hat{p}_0) dt \tag{9}$$

$$e_H[k] = H[k] - H[k-m], \quad e_{\hat{H}}[k] = \hat{H}[k] \tag{10}$$

$$J = \frac{1}{2}(e_H[k] - e_{\hat{H}}[k])^2 \tag{11}$$

$$\begin{bmatrix} \hat{p}_n[k] \\ \hat{p}_{n-1}[k] \\ \vdots \\ \hat{p}_0[k] \end{bmatrix} = \begin{bmatrix} \hat{p}_n[k] \\ \hat{p}_{n-1}[k] \\ \vdots \\ \hat{p}_0[k] \end{bmatrix} + \lambda \cdot (e_H[k] - e_{\hat{H}}[k]) \cdot \begin{bmatrix} w^n[k] \\ w^{n-1}[k] \\ \vdots \\ 1 \end{bmatrix} \tag{12}$$

SYSTEM IMPLEMENTATION AND TESTS

The developed system is integrated and fabricated modular units based on the circuit configuration. All the modules are available and ready to use. The implementation work needs to design suitable power supply; data interface and control circuit with accurate strategy to carry out the expected function capability. System software on data bus is programmed with appropriate protocol. Under the integrated concept, the proposed system will operate GPS data acquisition and INS correction to enhance the navigation performance. To accomplish the capability of VU, the odometry and gear-box speed are test and compared with GPS velocity in ARTC campus.

A. ODOMETRY AND GEAR-BOX SPEED TEST

The odometry hardware used frequency to voltage chip to convert signals into voltage. The dynamic test is compared with GPS velocity. The odometry speed test is implemented in the ARTC campus, and operator drove to the road terminal and turn left/right. Fig. 8 showed the variation and difference contrast to GPS speed. The frequency of odometry signal was varying from 0 Hz to 600 Hz, and then converted to voltage (0.0~5.0V) using charge-pump. In low speed operation, the odometry signal has large variation because of disturbance in Fig. 8(a)-(b). However, the mostly result is very similar to GPS velocity and this test is accomplished in calm weather. The Fig. 8(b) is special used to test availability and know how the available speed is. The odometry sensor has high feasibility and could be used in vehicle test when the speed is larger than 10kph. Although the odometry had shortcoming in low speed driving, the proposed system adopts gear-box speed as alternative solution in low speed. The gear-box speed is used to instead of odometry in low-speed, but higher speed is also adopted from odometry information because of turning angles. The gear-box signal is digital level, and its frequency varies from 0.5 Hz to hundred Hz with speed variation. Fig. 9 is the vehicle dynamic test comparing with GPS speed.

B. INERTIAL SENSOR CALIBRATION AND TEST

After INS had been set well in demonstrated vehicle, a driver drove in different speed to test straight moving and turn availability. The straight driving test is used to adjust accelerometer parameters refer to GPS speed. The output data would be processed using integration. Owing to integration error, the result should be calibrated and delicate processed well. Fig. 10(a) used a one-axis acceleration to get speed and the result is compared with GPS speed. The parameters was learned and calibrated by parameters learning and error cancelation. Hence, the speed error is under 5kph. In the similar way, the gyro integrated angular rate into heading comparing with GPS course w.r.t North direction in Fig. 10(b).

C. DISTANCE TEST IN AUTONOMOUS FUNCTION

This paper adopted a realization test in a car. In the integration of INS, the GPS provide a good position and heading. The reliability for long period is well, but it might be affect by the environment. The characteristic of INS is autonomous and reliable in short period. But the integration error is as larger as time. So the integration of GPS and IMU has a good advantage in compensating each other. The GPS can be an error bound of INS. Fig. 9 is the test area, and the related experiments are test in different driving procedures from reference points located in the terminal of the road. The test included static tests in Table 2 and dynamic result in Table 3.

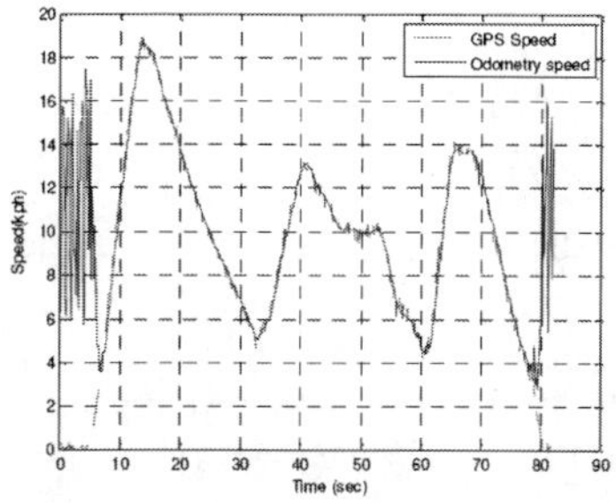

Figure 6(a). Driving test from 0~20kph.

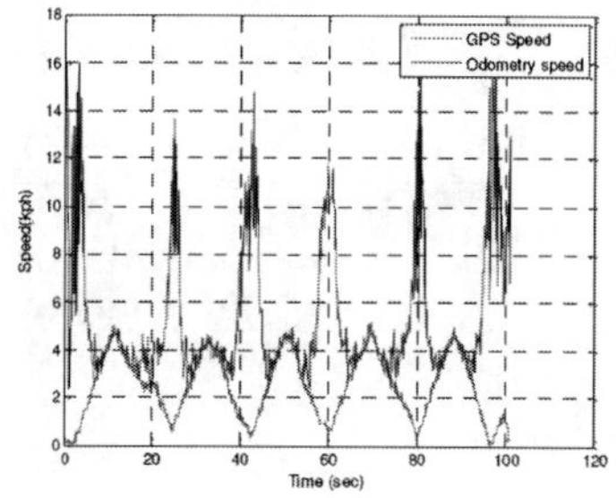

Figure 6(b). Driving test from 0~5kph.

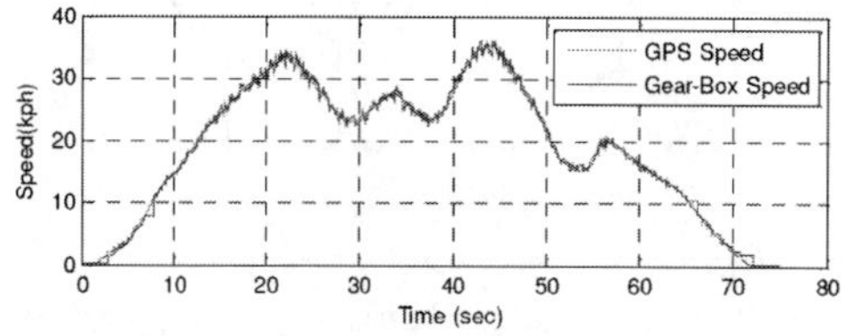

Figure 7(a). Gear-box speed vs GPS speed.

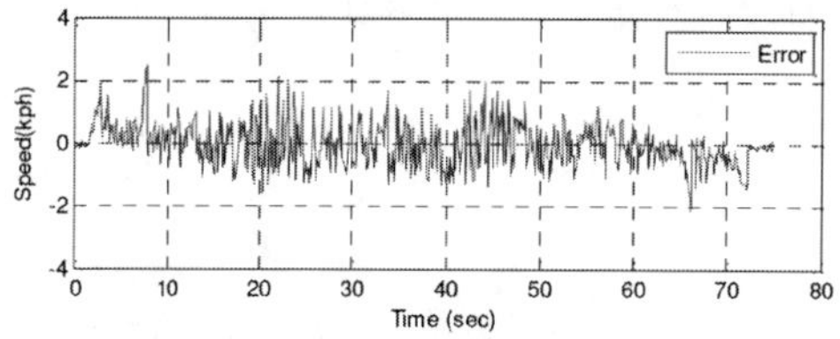

Figure 7(b). Speed error vs GPS speed.

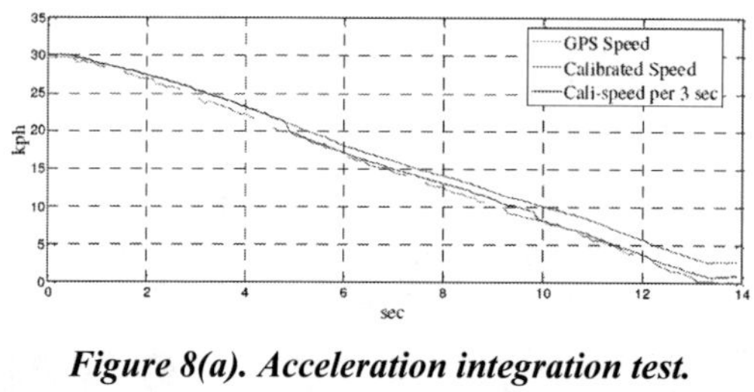

Figure 8(a). Acceleration integration test.

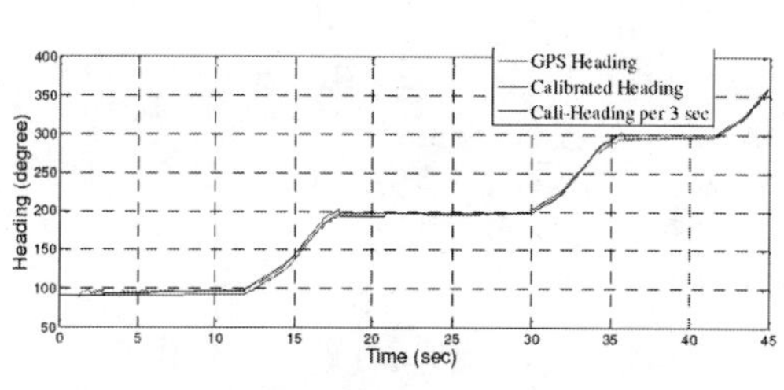

Figure 8(b). Gyro integration test.

Table 2. Static tests result.

Ref. Points / Error index	ARTC7	ARTC8	ARTC9	ARTC10	ARTC11	ARTC12	ARTC13	ARTC14
RSS (m)	10.097 10.097	7.062 7.062	7.592 7.666	6.487 6.486	5.351 5.350	4.904 7.998	8.856 8.320	4.200 4.200
North Error(m)	-7.140 -7.139	-4.994 -4.993	-5.364 -5.420	-4.587 -4.586	-3.783 -3.783	-3.468 3.952	-6.257 -5.871	-2.970 -2.970
East Error(m)	3.601 3.959	3.990 3.823	5.790 5.842	4.874 4.709	3.289 4.063	5.570 -1.981	3.950 3.950	5.458 5.458
CEP50 (m)	0.1718 0.158	0.144 0.1	0.295 0.126	0.145 0.1	0.341 0.158	0.115 6.952	0.338 0.451	0.1 0.1
CEP90 (m)	0.308 0.280	0.253 0.1	0.537 0.212	0.255 0.1	0.621 0.280	0.182 12.674	0.616 0.821	0.1 0.1

Table 3. Dynamic tests result.

Tests / Reference points	Test1	Test2	Test3	Test4	Average distance and its error rate	Max. error rate
Ref.12→Ref.11(N to S)	97.10	97.03	97.47	95.95	96.89	
•RTK 97.8 m	0.72%	0.79%	0.34%	1.89%	0.93%	
•RMS 96.4 m	0.73%	0.65%	1.1%	0.47%	0.74%	
Ref.10→Ref.9(W to E)	224.23	223.06	224.53	223.24	223.77	
•RTK 221.9 m	1.05%	0.52%	1.19%	0.60%	0.84%	
•RMS 221.2 m	1.37%	0.84%	1.51%	0.92%	1.16%	3.32%
Ref.8→Ref.7(S to N)	93.92	95.04	92.72	92.72	98.5	3.22%
•RTK 95.9 m	2.06%	0.90%	3.32%	3.32%	2.4%	
•RMS 95.8 m	1.96%	0.79%	3.22%	3.22%	2.3%	
Ref.14→Ref.13(E to W)	224.27	223.47	223.24		223.66	
•RTK 222.7 m	0.70%	0.35%	0.24%		0.43%	
•RMS 225.8 m	0.68%	1.03%	1.13%		0.95%	

Table 4. Dynamic tests result.

Position (m)	Packet Loss Rate (%)	Latency (msec)
+200	76	654.8
+100	70	461.7
0	55	19.6
−100	66	668.6
−200	54	361.3

Figure 9. Test field in ARTC campus.

D. CAR-FOLLOWING INTEGRATION TEST

Under system implementation, the platform static tests include packet loss and latency. The environment noise was measured before static test in Fig. 10. The power of environment noise was −115 to −95 dbm, and it was too small to affect the packet tests. Each result would minus 20 because of probe decay. The range test showed in Table 4, and the packet loss rate rule is to get 95% successful receiving at 1000 times packet transmitting.

After the static tests were finished, the dynamic test of cooperative driving was operated. The test platform, as shown in Fig. 11, was used to implement the proposed car-following function. Three vehicles were used to run on the ARTC roadway to verify the geocasting and neighboring situation function; while three vehicles were followed on roadway or driven to an intersection in order to easily verify system design in the verification. Under system design and implementation, vehicular data which connected to CAN interface was collected for scheduled tests. For test operations, DSRC module was used to exchange positioning

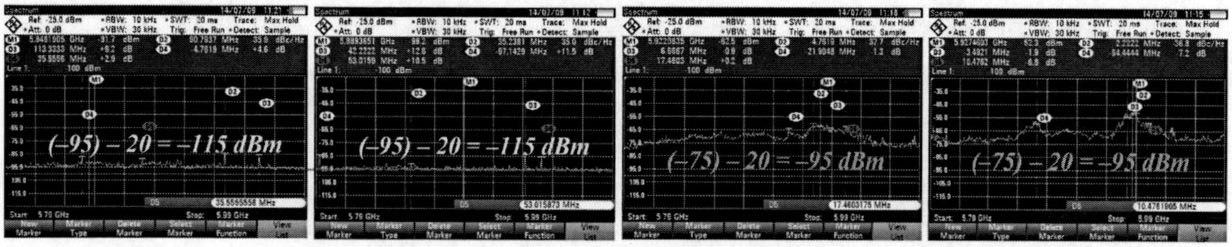

Figure 10. Background measurement using frequency spectrum.

Figure 11. System Hardware and Implementation.

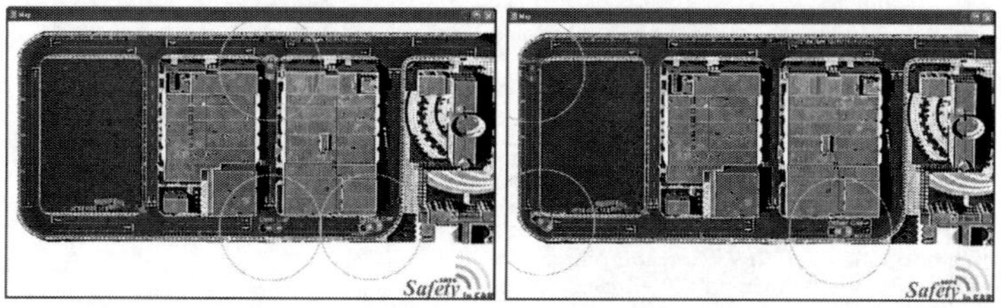

Figure 12. Inter-vehicle geocasting result in intersection

data. The INS module also outputted positioning data, and the positioning performance was well processed and mapped to demonstrated map. Fig. 12 showed the inter-vehicle geocasting tests, where two cases showed that three vehicles were individually running on ARTC roadway. Two tests demonstrated the car-following away from other vehicles, and the driver drove far away from preceding vehicle or following the other vehicle. Fig. 13 showed another test, where the driver drove into an intersection. These cases offered an important awareness to the driver under test. The map reported neighboring vehicles position periodically, and broadcasted to other vehicles via DSRC communication. In these tests, the actual position of the vehicle was monitored and displayed on the screen. The system concept provided a car following that the operator could be fully aware of 360-degree vehicles situation under system concept.

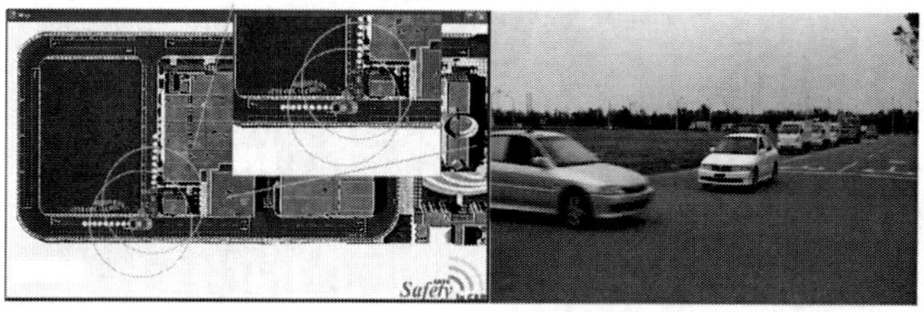

Figure 13. Car-following in blind intersection

CONCLUSIONS

In this paper, the proposed concept demonstrates a cooperative driving or car-following concept for remote surveillance applications. The system design simplified using embedded microprocessor with DSRC module to activate UDP and port 802.lip protocol into data link, continuously vehicular speed using CAN network and high sensitivity INS positioning. The proposed system assists drivers to know current relationship to other vehicles through intersection and following tests with the situation awareness capability. Although the result of DSRC packet loss rate and latency are not very good, this drawback will be improved in progress concern. The demonstrated tests have been verified the availability of collision estimation and inter-vehicle geocasting algorithm.

Several hardware, software and firmware are implemented to realize the concept of autonomous navigation. Combining with the GPS/IMU integration, the vehicular signals could play a potential auxiliary support to derive moving information to enhance capability. The advantage of autonomous system has presented fewer than 4% position error, and the demonstration provides a higher availability solution for vehicle position to enhance cooperative driving.

REFERENCES

1. "Death statistics from 1999 to 2010 of road traffic safety portal transportation in Taiwan". Available at http://168.motc.gov.tw/GIPSite/wSite/mp
2. Xiaohong, (Allison), Zhang (M.Sc. thesis), "Integration of GPS with a medium accuracy IMU for metre-level positioning," *UCGE Report No. 20178*, June 2003.
3. Sengupta, R., Rezaei, S., and Shladover, S. E. et al., "Cooperative Collision Warning Systems: Concept Definition and Experimental Implementation," *Journal of Intelligent Transportation Systems*, Volume 11, Issue:3, July 2007, pages 143-155.
4. Jiang, D., Taliwal, V., Meier, A., "Design of 5.9 GHz DSRC-Based Vehicular Safety Communication," IEEE Transactions on Wireless Communications, Oct. 2006, Vol. 13, Issue: 5, pp. 36-43.
5. Lin, C. E., Hsu, C. W., and Lee, Y. S. et al., "Verification of UAV Flight Control and Surveillance using Mobile Communication," AIAA Journal of Aerospace Computing, Information and Communication, Vol. 1, No. 4, April 2004, pp. 189-197.
6. Fernandes, P., and Nunes, U., "Vehicle Communications : A Short Survey," IADIS Telecommunications, Networks and Systems, Lisbon, 2007, pp. 134-138.
7. Hwang, D. H., Moon, S. W., Sung, T. K., Lee, S. J., "Design and implementation of an efficient tightly-coupled GPS/INS integration scheme", Proceedings of 2000 National Technical Meeting, 2000, pp. 159-165.
8. Zuberl, K. M., and Shin, K. G., "Scheduling Messages on Controller Area Network for Real-Time CIM Applications," IEEE Transactions on Robotics and Automation, Vol. 13, No. 2, pp.310-314, April 1997.
9. ISO 9141: Road Vehicles - Diagnostic System - Requirements for Interchange of Digital information, International Standard Organization, Oct. 1989.
10. Stojmenovic, I., "Geocasting with Guaranteed Delivery in Sensor Networks," IEEE Wireless Communications, Vol. 11, Issue:6, Dec. 2004, pp. 29-37.
11. Chen, C.T., Tekinay, S., and Papavassiliou, S., "Geocasting in Cellular Ad hoc Augmented Networks," Vehicular Technology Conference, Oct. 2003, Vol. 3, pp.1858-1862.

CONTACT INFORMATION

Chan Wei Hsu
winsonhsu@artc.org.tw
Tel:+886-4-7811222#2351
Fax:+886-4-7811333
Address: No.6, Lugong S. 7th Rd., Lugang Town, Changhua County 50544, Taiwan (R.O.C.)
http://www.artc.org.tw/index_en.aspx

ACKNOWLEDGMENTS

This work is supported in research projects 100-EC-17-A-04-02-0803 by Department of Industrial Technology, Ministry of Economic Affairs, Taiwan, R.O.C.

ABBREVIATIONS

GPS
global positioning system abbreviations

EMU
inertial measurement unit

INS
inertial navigation system

DSRC
dedicated short range communication

VU
vehicle unit

CAN
controller area network

UART
universal asynchronous receiver and transmitter

OBD
on-board diagnostic

CEP
circular error probable

RMS
root mean square

SAE International
2012-01-0491
Published 04/16/2012

doi:10.4271/2012-01-0491
saepcelec.saejournals.org

DSRC Performance Comparison with and without Antenna Diversity Using Different Transmission Power

Sue Bai and Radovan Miucic
Honda R&D Americas, Inc.

ABSTRACT

Vehicle-to-Vehicle (V2V) safety application research based on short range real-time communication has been researched for over a decade. Examples of V2V applications include Electronic Emergency Brake Light, Do Not Pass Warning, Lane Departure Warning, and Intersection Movement Assist. It is hoped that these applications, whether present as warning or intervention, will help reduce the incidence of traffic collisions, fatalities, injuries, and property damage. The safety benefits of these applications will likely depend on many factors, such as usability, market penetration, driver acceptance, and reliability. Some applications, such as DNPW and IMA, require a longer communication range to be effective. In addition, Dedicated Short Range Communications (DSRC) may be required to communicate without direct line of sight. The signal needs to overcome shadowing effects of other vehicles and buildings that are in the way. Some remedies may include increasing transmission power, using higher gain antennas, and using multiple antennas with diversity reception. This paper presents experimental results of several field trials using different configurations in transmission power, antenna diversity, and antenna gain of a DSRC receiving system.

CITATION: Bai, S. and Miucic, R., "DSRC Performance Comparison with and without Antenna Diversity Using Different Transmission Power," *SAE Int. J. Passeng. Cars - Electron. Electr. Syst.* 5(2):2012, doi:10.4271/2012-01-0491.

INTRODUCTION

Dedicated Short Range Communication (DSRC) is an enabling technology for wireless cooperative safety systems between vehicles. The wireless physical layer is defined by the IEEE 802.11p [1] amendment to 802.11 [2] standard. Upper layers are captured in IEEE 1609.3 [3] and 1609.4 [4]. Security is defined in IEEE 1609.2 [5] Safety-related payload content is defined in SAE J2735 [6].

Broadcast beckons called Basic Safety Messages (BSMs) are being sent out periodically from the equipped vehicles. A BSM contains vehicle information such as speed, location, acceleration, brake status, and yaw rate. By exchanging this basic information, vehicles are becoming more aware of potential hazards and may be able to avoid or mitigate collisions.

This paper shows recent experimental measurements of DSRC performance on a typical suburban blind intersection using single and dual antenna setups, different transmission power levels, two antennas with different gains, and two different types of receivers. This paper analyzes the transmitted power and communication distance and how they influence cooperative safety applications.

BODY

OBJECTIVE

V2V communication without line-of-sight (LOS) truly demonstrates the advantage of wireless-based cooperative safety systems compared to stand-alone autonomous sensors such as radars and lidars. For cooperative safety systems to be effective, vehicles need to have situational awareness and communicate in non-LOS scenarios. Likewise, a vehicle hosting an Intersection Movement Assist (IMA) application needs to communicate to the vehicles coming from a blind intersection and alert the driver of a potential collision so that he can react and respond appropriately.

Increasing power or using antenna diversity overcomes communication issues in non-line-of-sight scenarios. The study can be used to support fine-tuning of network simulators. Also, the results offer some real-world experimental insights about the power needed for a required range of communication to support wide variety of intersection-related safety applications. The presented communication performance in the intersection scenario can be used in future research to configure meaningful simulations that investigate the radio channel characteristics at intersections. The research quantifies the relative improvements in performance due to power increase, higher antenna gain, packet switching diversity, and antenna diversity with respect to each other.

RELATED WORK

This paper shows communication performance in terms of packet success ratio and the experimental relationship between range and power for the different communication configuration for a non-line-of-sight intersection. To date, there is no documented research of a comprehensive study of a blind intersection with different configurations.

In terms of related work, Mangel [12] analyzed the characteristics of various intersections in a city area and provides the basis of the needs for researching DSRC reception at Non-Line-Of-Sight environment. Alexander [7] describes performance of two radio chip set receivers in a closed intersection scenario; however, the work looks only at a single transmitted power value. Hong [8] shows empirical results, in terms of PER, RSSI and T-window [9] analysis, of field testing using 10, 20, and 33 dBm transmission power in urban and suburban intersections. In addition to analyzing power versus range relationship, this paper examines effects of different radio chipsets, antenna diversity, and antenna gains at a suburban intersection. Mangel [11] compared the reception rate and reception power using a fixed-location transmitter at various intersection points using a single antenna. While the test provides a general performance of the specific setup, it did not consider the potential of using antenna diversity to improve the reception performance, which can be substantial in a similar environment with a strong multi-path effect.

TEST SETUP

The tests were conducted using two vehicles. One served as the receiver and the other as the transmitter. Two types of Onboard Equipments (OBEs), referred to in this paper as Type A and Type B, were used in the test vehicle. Both OBEs have internal memory for logging the over-the-air transmitted and received messages. Type A OBE uses a proprietary radio chipset while Type B uses a commercial Atheros chip.

TRANSMITTER SETUP

As shown in Fig. 1, the transmitter vehicle was equipped with the Type A DSRC OBE, NovaTel Global Positioning System, and MobileMark omni-directional antenna. This transmitter setup was kept the same during the tests. The transmitter sent periodic DSRC BSMs every 100 ms. Transmission power was changed from test to test, but it remained a constant during execution of a particular test.

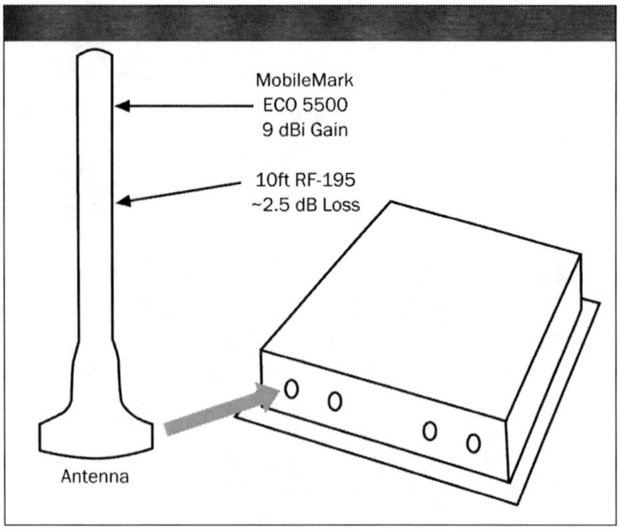

Figure 1. Transmitter vehicle setup: Type A DSRC radio module and MobileMark omni-directional antenna. This configuration was used as transmitter vehicle setup for all tests.

RECEIVER CONFIGURATION

The receiver vehicle setup configuration was changed per different testing needs. The receiver vehicle was instrumented with several configurations consisting of different combination of the DSRC radio types, and antennas. In the first setup configuration the receiver vehicle had two omni-directional high gain (9 dBi on horizon) antennas mounted on the roof rack. These antennas do not require a dedicated ground plane and therefore being on the rack mounting fixture did not impact performance. Signal routed via RF-195 cable to the trunk of the vehicle where the rest of the equipment resided. Due to the cable loss signal was attenuated by about 2.5 dB.

Furthermore, the signal from each antenna was split into two feeds via signal dividers. Again the signal was attenuated by an additional 3 dB. Finally, a low loss cable LMR-400 feed the signals to the OBEs. The receiver vehicle Configuration 1 is shown in Fig. 2. This configuration is intended to capture over-the-air messages received in Type A OBE with antenna diversity functionality. Also this configuration simultaneously captures messages in the type B OBEs without diversity. As shown in Fig. 3, Configuration 2 is very similar to Configuration 1 except the antennas are now Nippon omni-directional antennas with lower gain (0 dBi on horizon). The Nippon antenna requires a ground plane, so it was magnetically mounted on the vehicle's roof. Configuration 3 consists of two pairs of Type A and Type B OBEs, that is, one Nippon antenna and one MobileMark antenna.

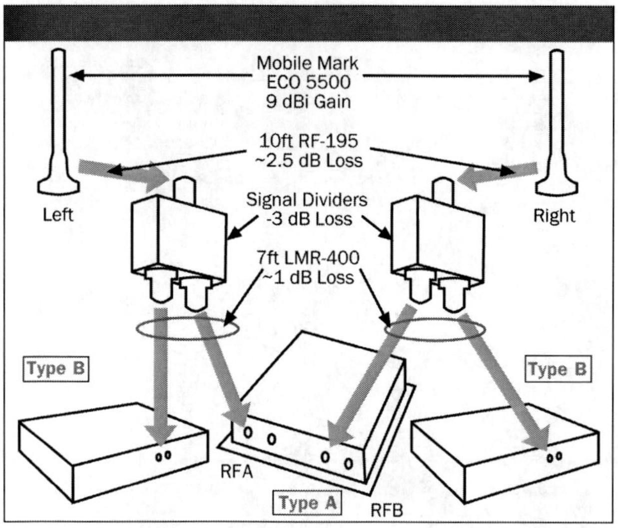

Figure 2. Receiver vehicle Configuration 1 consisting of two MobileMark antennas, two Type B OBEs, and one Type A OBE.

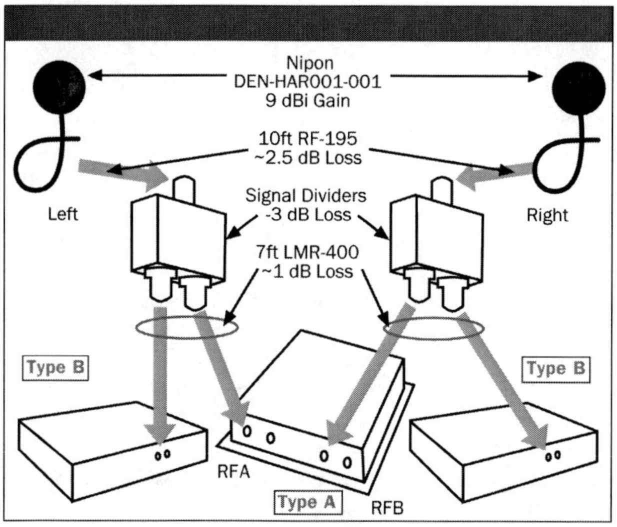

Figure 3. Receiver vehicle Configuration 2 consisting of two Nippon antennas, two type B OBEs, and one Type A OBE.

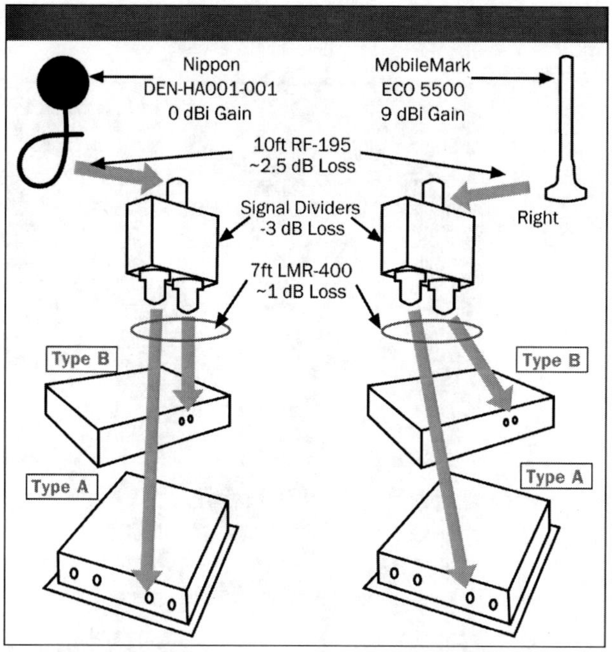

Figure 4. Receiver vehicle Configuration 3 consisting of one Nippon antenna, one MobileMark antenna, two type B OBEs, and two Type A OBEs.

TEST SCENARIO

As shown in Fig. 5, a typical suburban street was chosen to represent a closed intersection environment. A receiver or application host vehicle (HV) was static and positioned 35 m from the intersection while the transmitter or remote vehicle (RV) was traversing.

Three experiments were conducted for each of the receiver vehicle configurations. The power of the transmitter was varied from 8 to 25 dBm at increments of 2 dBm with exception of the test with 25 dBm (maximum power type A OBE could produce). Each of the power levels were tested with 5 laps. Overall, the transmitter vehicle traveled 150 laps over several days of testing. The vehicle traveled up to 25 mph.

DATA RESULTS AND ANALYSIS

Data was logged on the internal OBE memory and was off-loaded to a test laptop after each test. Based on difficulties in obtaining data from related research [10], redundant UDP packet data collection was used in addition to the data written to the flash memory. As a risk management measure, special software was written to output UDP messages from each OBE every time a DSRC message is sent or received. The OBEs were assigned a static IP address and connected via a LAN router to the test laptop. The UDP traffic was then logged to the test laptop. While driving in areas of heavy foliage from trees, the GPS often did not provide a solution as evidenced by missing positioning points

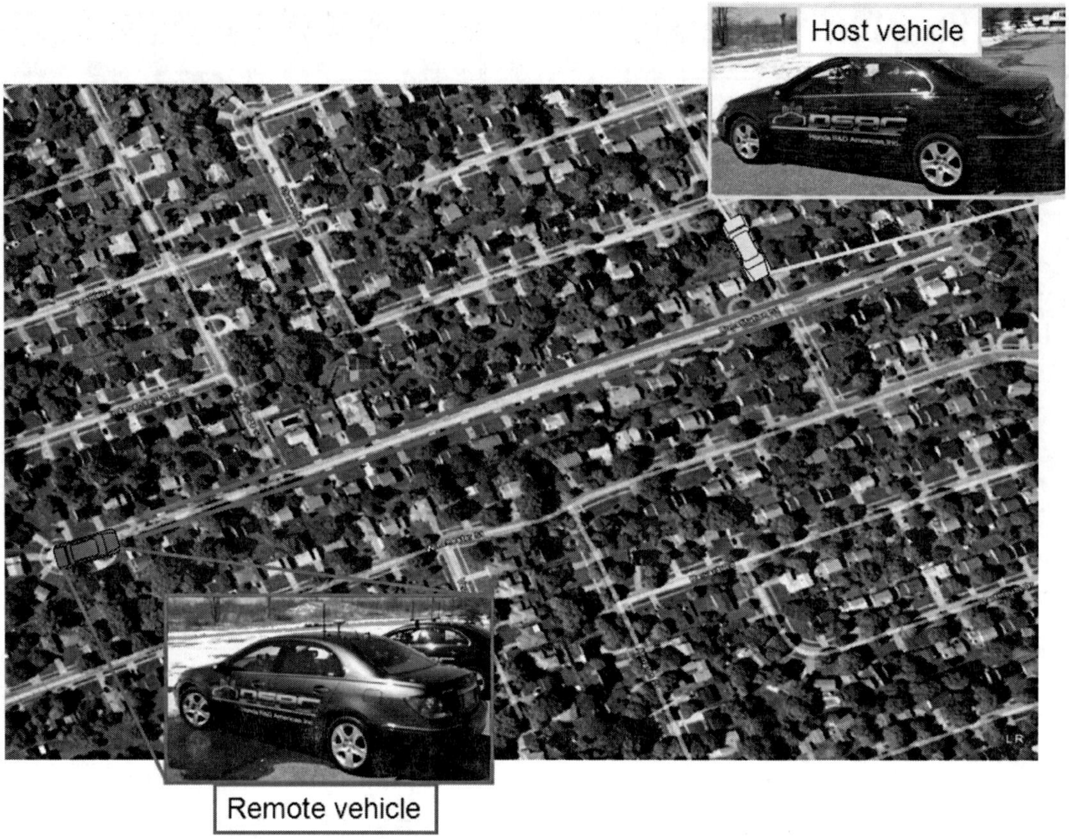

Figure 5. Aerial view of the testing site.

in the log. Therefore, during post processing, the missing points were added to the log by interpolating known previous and future position points. Some error was introduced, but the amount was insignificant for this study.

PERFORMANCE METRIC

Performance was measured in terms of packet success ratio (PSR), which measures a receiver's ability to successfully receive a packet sent from the transmitter. Rolling sequence numbers are part of the BSM. The PSR is calculated by comparing sequence numbers of the transmitted packets to the sequence numbers of the received packets. Results are divided into 5 m distance-bins and plotted on a graph as PSR (%) versus distance (m). Since the transmitter vehicle traveled no more than 25 mph, there are 10 data points on average per bin per lap. So for one test, given a particular receiver configuration and power level, there are on average, 50 data points per bin. Furthermore, results are categorized as incoming if the transmitter was approaching the receiver and outgoing if the transmitter was moving away from the receiver. This was done to see if there was any significant bias due to the direction of the vehicles.

Test 1 and Test 2

The Type A receiver uses maximal-ratio combining diversity to add the two signals at the receiver and encode the packet. The incoming and outgoing results are shown in Fig. 6. In this research, the Type B receiver was not enabled for antenna diversity reception.

However, in the Configuration 1 and Configuration 2 setups, there are two Type B receivers. Each receiver was connected to a single antenna and received a signal from a single feed. In post processing, logs were combined from both receivers. A successful packet reception is declared when a particular packet is received in either or both receivers. This method is referred to as packet switching diversity.

Figure 7 shows PSR performance for Type B OBE reception.

Table 1 shows that combining the maximum ratios improves the communication range from 3 m up to 75 m compared to packet switching diversity. The mean improvement is 23.78 m with standard deviation of 17.4 m. There are no significant differences between incoming and outgoing results. Figures 8 and 9 show the communication range and improvements. The setups in Configurations 1 and 2 split the power of the antennas into two feeds. The power that reaches Type A OBE comes from both antennas. The power of the antenna signal that reaches type B receiver is from a single feed.

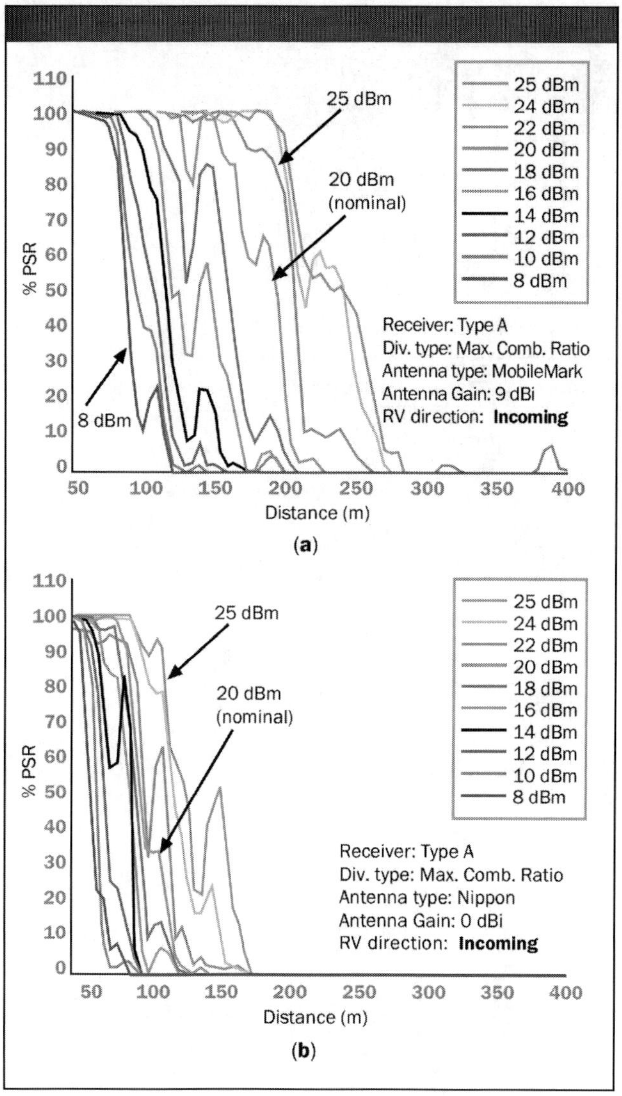

Figure 6. PSR versus distance for Type A receiver with antenna diversity (Maximum Ratio Combining) using a) 9 dBi and b) 0 dBi antennas.

Figure 10 shows the difference in range of the two antennas for the Configuration 1 and Configuration 2 setups. The figure shows communication range improvement at higher transmission powers for the Maximum Ratio Combining compared to the Packet Switching diversity scheme. However, the range improvement at the higher transmission powers appears to be proportional to the improvements at the lower transmission powers.

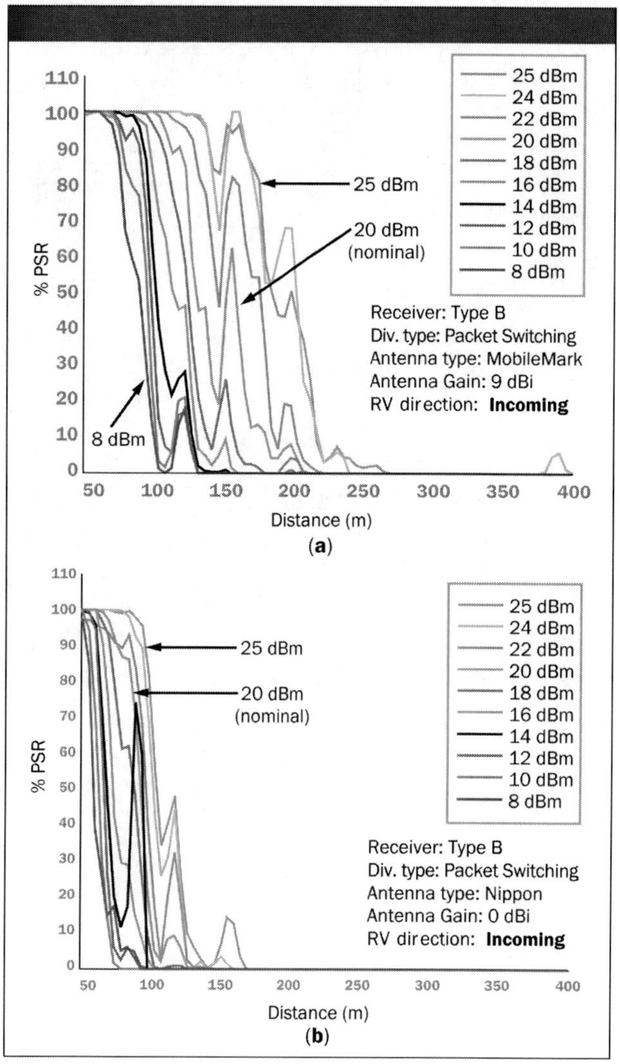

Figure 7. PSR versus distance for type B receiver with antenna diversity (Packet Switching) using a) 9 dBi and b) 0 dBi antennas.

Test 3

Test 3 uses Configuration 3 setup for the transmitter vehicle. Type A OBE uses advanced channel tracking reception, while the Type B OBE uses a commercial method based on a slightly modified WiFi chipset. The objective is to measure this difference given the same non-line-of sight wireless environment and no antenna diversity.

Figures 11 and 12 show communication range performances at varying transmission power for the Type A and Type B OBE.

Table 1. Communication range (in m) for the Maximum Ratio Combining and Packet Switching antenna diversity schemes for incoming and outgoing direction of the transmitter path. Table shows vehicle separation distances for which PSR drops below 70% and 90%.

Incoming												
Ant.	0 dBi						9 dBi					
PSR	70%			90%			70%			90%		
OBE	A	B	A-B	A	B	A-B	A	B	A-B	A	B	A-B
Power [dBm] 8	46.5	42.5	4	42	39	3	84.5	65	19.5	79	60	19
10	52.5	46	6.5	48	42	6	88	78	10	82	63.5	18.5
12	57.5	49.5	8	51.5	46.5	5	97.5	81.5	16	87	76	11
14	62	51	11	56	47	9	112.5	84.5	28	99	79	20
16	75	56.5	18.5	57.5	51	6.5	116	92	24	110	83.5	26.5
18	82	63.5	18.5	75.5	57	18.5	128	105.5	22.5	117	93.5	23.5
20	86.5	76	10.5	81	62	19	170	114	56	129	99.5	29.5
22	88.5	81	7.5	81.5	63	18.5	203	128	75	175	117	58
24	108.5	86	22.5	89	79.5	9.5	206.5	135	71.5	198	128	70
25	110	88.5	21.5	93	83	10	210	168.5	41.5	202	128.5	73.5

Outgoing												
Ant.	0 dBi						9 dBi					
PSR	70%			90%			70%			90%		
OBE	A	B	A-B	A	B	A-B	A	B	A-B	A	B	A-B
Power [dBm] 8	48	44.5	3.5	43	39.5	3.5	83.5	72	11.5	78.5	58	20.5
10	56	46.5	9.5	50	42.5	7.5	90	77	13	83.5	62.5	21
12	62.5	49	13.5	54.5	43	11.5	107	83	24	87.5	78	9.5
14	78	50	28	59	44	15	112	86.5	25.5	106	80.5	25.5
16	81.5	60	21.5	64	52.5	11.5	117	99.5	17.5	111.5	84	27.5
18	84.5	65.5	19	77.5	58.5	19	161	111.5	49.5	118	90.5	27.5
20	88.5	80	8.5	83	64	19	170	117	53	162.5	111.5	51
22	107	83	24	83	65.5	17.5	184	159	25	176	120	56
24	113.5	87	26.5	104.5	82.5	22	207.5	163.5	44	202	158	44
25	122	106.5	15.5	109.5	86.5	23	212.5	167.5	45	205.5	159.5	46

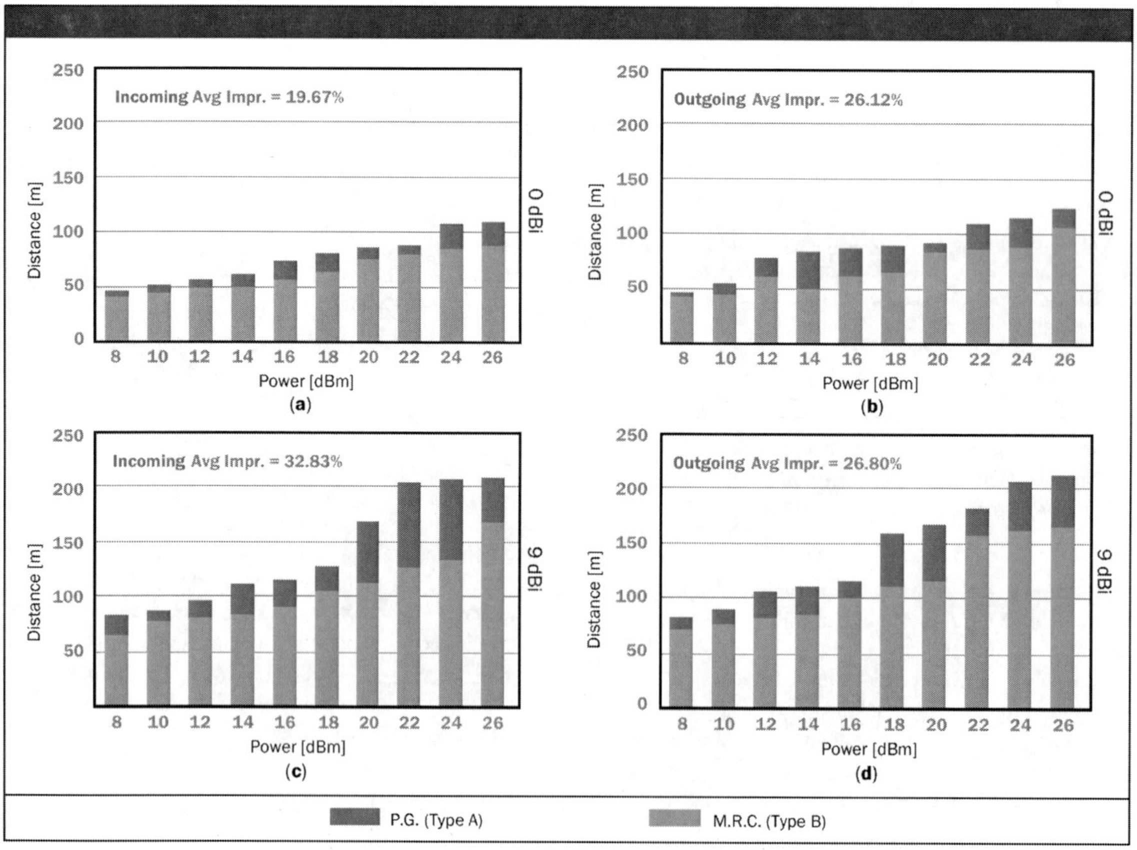

Figure 8. Test 1 and Test 2 (antenna diversity) communication range for the PSR of 70% and higher. The bottom two graphs are the results of the testing using 9 dBi antennas and the top two graphs are the results of the testing using 0 dBi antennas. The left two graphs are results of the incoming direction of the transmitter and right graphs are results of outgoing direction.

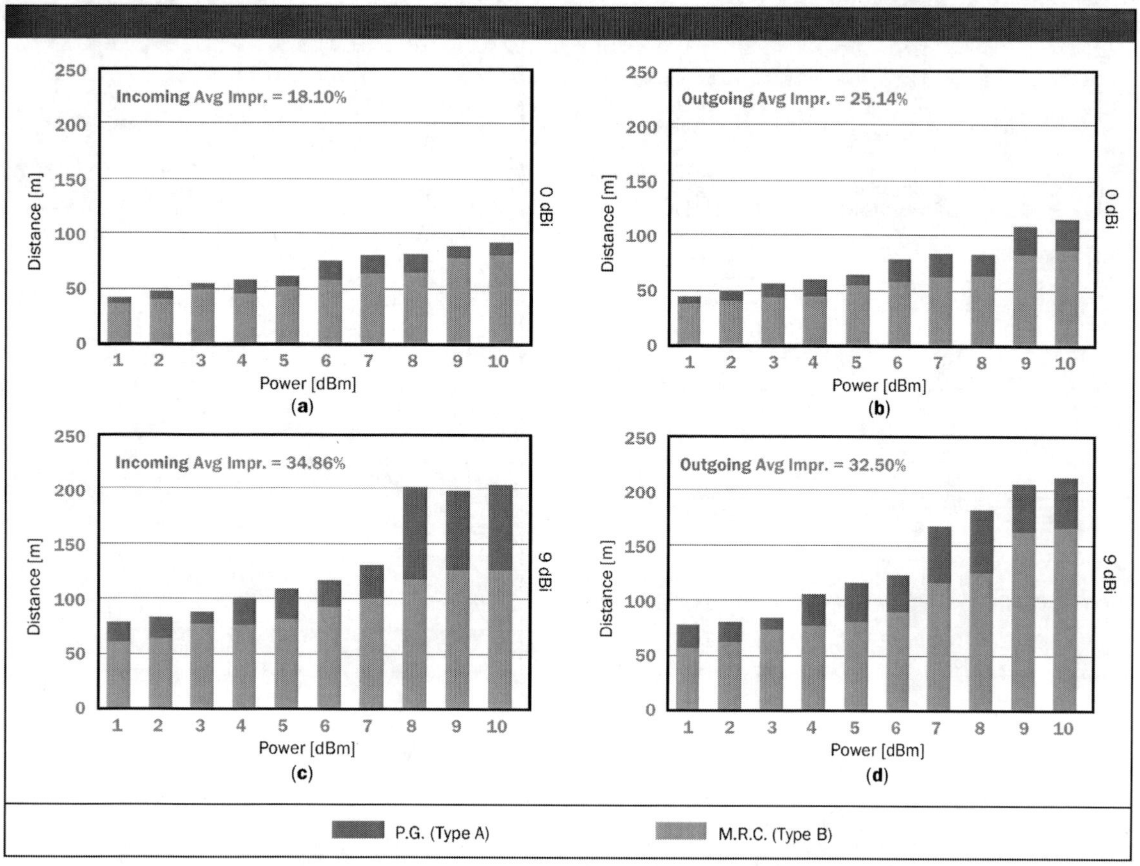

Figure 9. Test 1 and Test 2 (antenna diversity) communication range for the PSR of 90% and higher. The bottom two graphs are the results of the testing using 9 dBi antennas and the top two graphs are the results of the testing using 0 dBi antennas. The left two graphs are results of the incoming direction of the transmitter and the right graphs are results of outgoing direction.

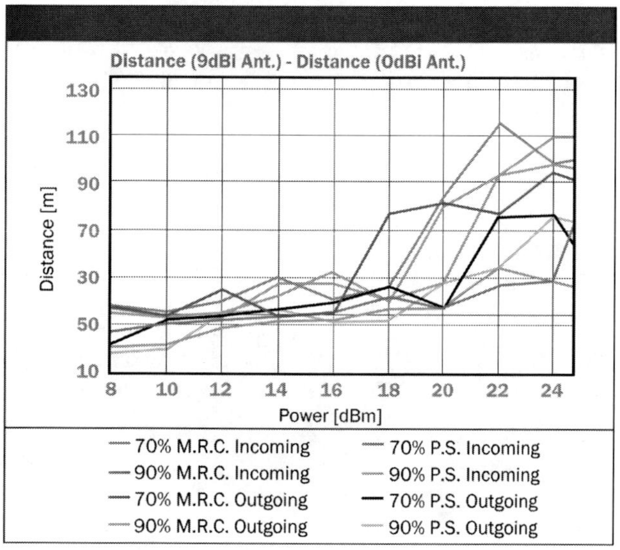

Figure 10. Comparison in range improvements between the 9 dBi and 0 dBi antennas for Test 1 and 2.

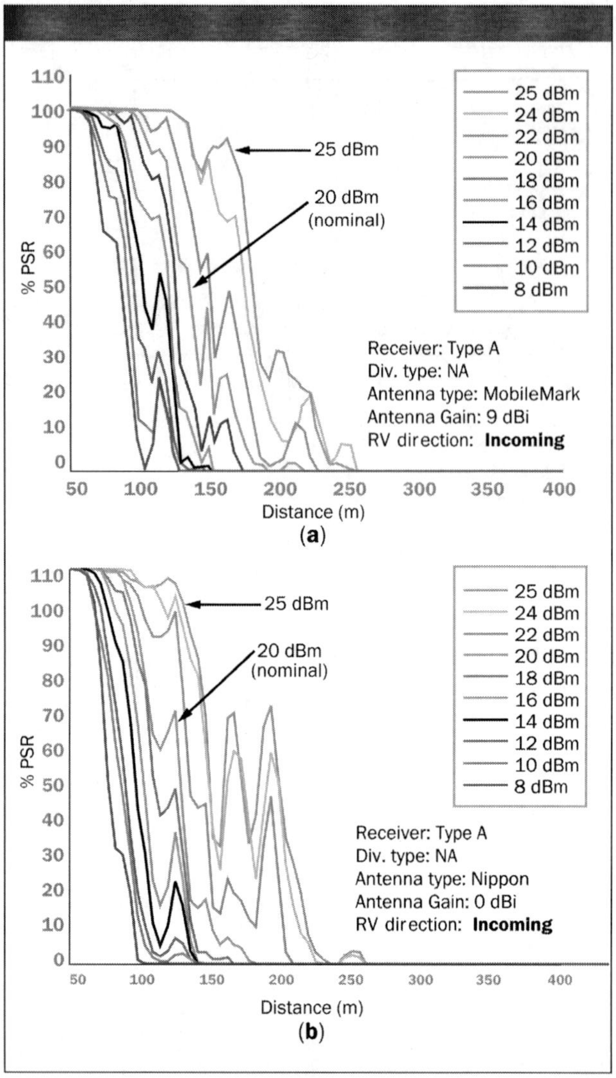

Figure 11. PSR versus distance for Type A receiver without antenna diversity using a) 9 dBi and b) 0 dBi antennas.

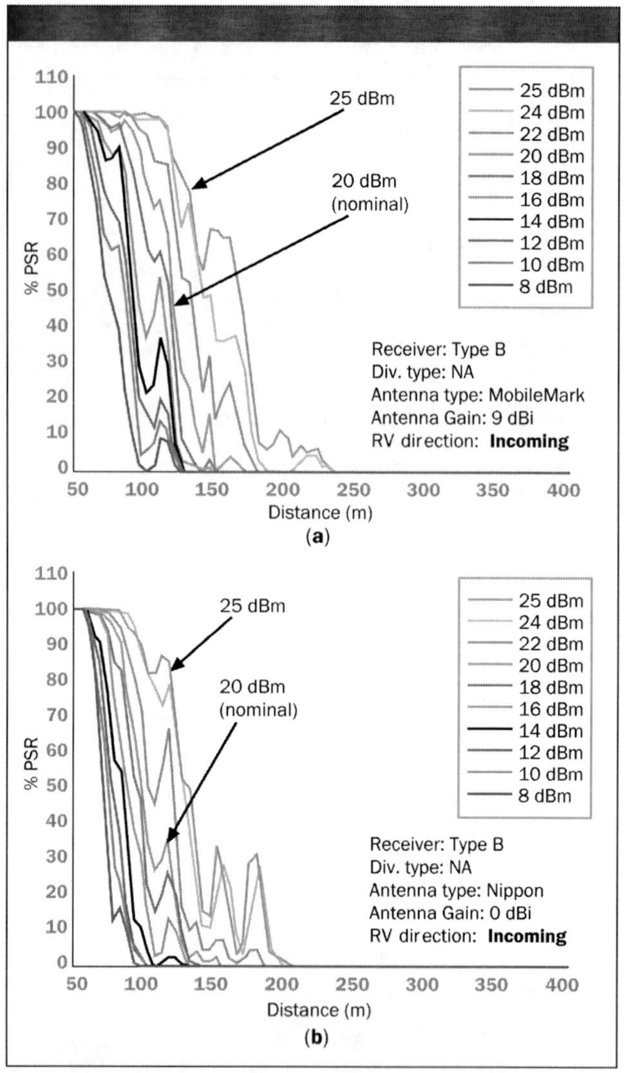

Figure 12. PSR versus distance for type B receiver without antenna diversity using a) 9 dBi and b) 0 dBi antennas.

Table 2 shows vehicle separation distances for which PSR is higher than 70% or 90%. As shown, Type A OBE communication range is from 1 m to 42 m longer than Type B OBE. The mean value is 10.7 m and standard deviation is 9.2 m. Figures 13 and 14 show the communication ranges and improvements.

To better understand the difference in performance of the two receivers without antenna diversity and using different antennas, the difference in communication range between the OBE types was analyzed. Figure 15 shows the difference in communication range for the given PSR, direction of transmitter travel, and OBE type. As shown, the communication range improvement was 3 m to 45 m when the 9 dBi antenna was used.

Table 2. Communication range (in m) without using diversity schemes, for incoming and outgoing direction of the transmitter path. Table shows vehicle separation distances for which PSR drops below 70% and 90%.

Incoming													
Ant.		0 dBi						9 dBi					
PSR		70%			90%			70%			90%		
OBE		A	B	A-B	A	B	A-B	A	B	A-B	A	B	A-B
Power [dBm]	8	57	56	1	51	50	1	65	62	3	58.5	54	4.5
	10	62	57.5	4.5	51.5	50	1.5	78.5	65	13.5	62	56.5	5.5
	12	64	60	4	56	52	4	83	75.5	7.5	64	59.5	4.5
	14	73	63	10	60.5	56	4.5	87.5	82.5	5	79.5	64	15.5
	16	77.5	69	8.5	65.5	62	3.5	99.5	84.5	15	82	68	14
	18	86	76	10	74	64	10	112.5	92.5	20	92.5	82	10.5
	20	86.5	79	7.5	77	71.5	5.5	116	109.5	6.5	98.5	90.5	8
	22	111	86.5	24.5	83.5	75	8.5	129.5	116	13.5	117	98.5	18.5
	24	122	110	12	98.5	84	14.5	154	120	34	122	113	9
	25	123.5	110.5	13	112	85	27	168.5	128.5	40	129	114	15

Outgoing													
Ant.		0 dBi						9 dBi					
PSR		70%			90%			70%			90%		
OBE		A	B	A-B	A	B	A-B	A	B	A-B	A	B	A-B
Power [dBm]	8	52.5	51	1.5	45	43	2	58.5	57	1.5	53	52	1
	10	57	54.5	2.5	50.5	46	4.5	78.5	63	15.5	58.5	56.5	2
	12	59.5	56	3.5	52	46.5	5.5	82	77.5	4.5	60	56.5	3.5
	14	67	59	8	57	54	3	85.5	82.5	3	78.5	57	21.5
	16	76	65.5	10.5	62.5	56.5	6	92	85	7	83	77.5	5.5
	18	81.5	76	5.5	72.5	58.5	14	110	85.5	24.5	82	79.5	2.5
	20	103	81	22	80.5	70	10.5	116	111	5	108	87.5	20.5
	22	111.5	90	21.5	88	83	5	125	115.5	9.5	116	110.5	5.5
	24	118	108	10	107.5	84	23.5	162	120	42	119	112	7
	25	122.5	109	13.5	109	86.5	22.5	163	123	40	120	112.5	7.5

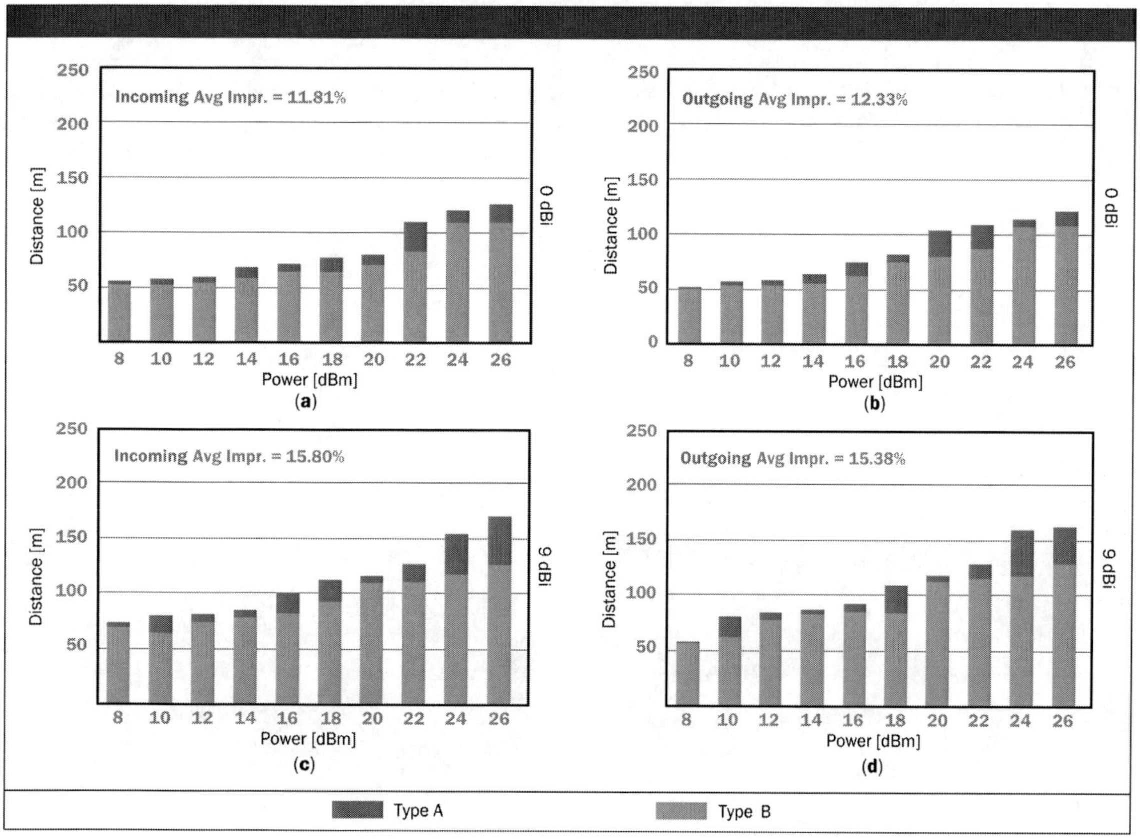

Figure 13. Test 3 (no antenna diversity) communication range for the PSR of 70% and higher. The bottom two graphs are the results of the testing using 9 dBi antennas and the top two graphs are the results of the testing using 0 dBi antennas. The left two graphs are results of the incoming direction of the transmitter and the right graphs are results of the outgoing direction.

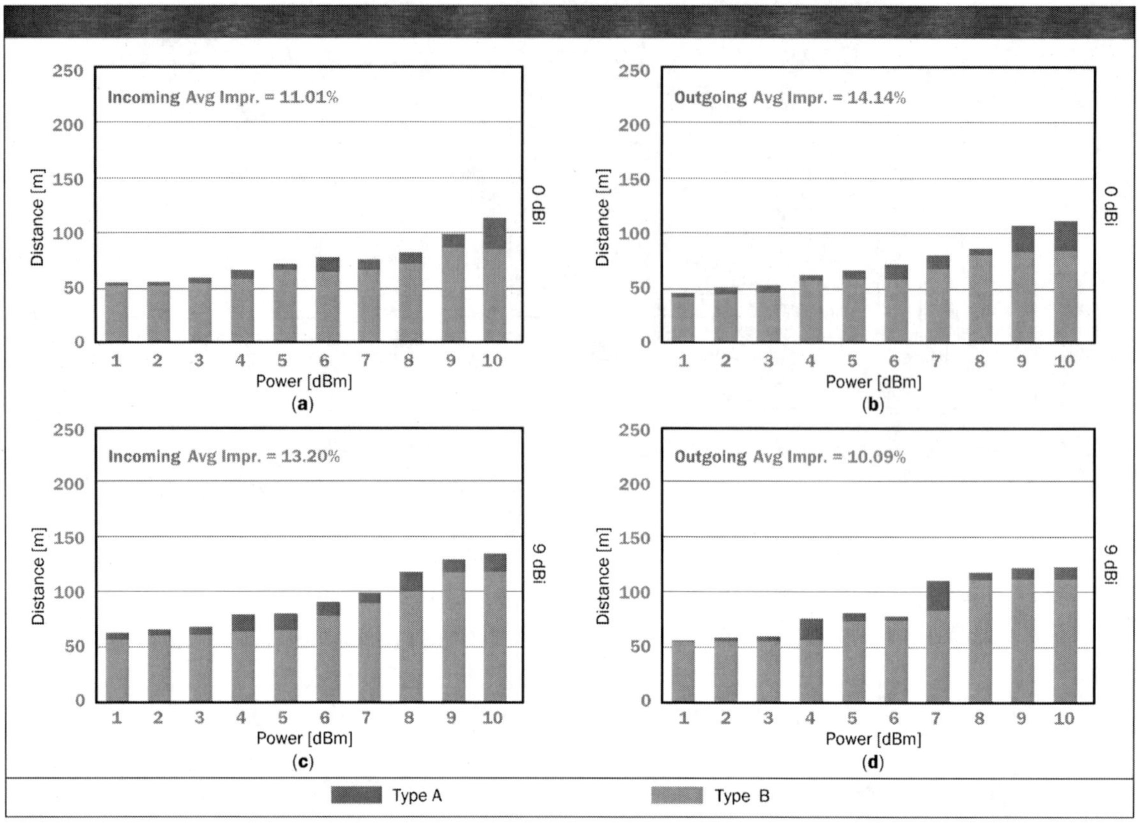

Figure 14. Test 3 (no antenna diversity) communication range for the PSR of 90% and higher. The bottom two graphs are the results of the testing using 9 dBi antennas and the top two graphs are the results of the testing using 0 dBi antennas. The left two graphs are results of the incoming direction of the transmitter and right graphs are results of the outgoing direction.

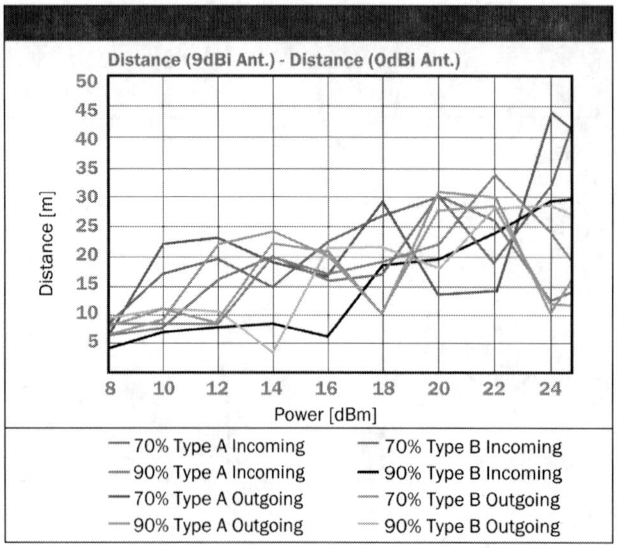

Figure 15. Test 3: range improvements comparison between the 9 dBi and 0 dBi antennas.

SUMMARY/CONCLUSIONS

This paper analyzes the DSRC communication reception performance improvement in NLOS scenarios using various configurations. The basis for comparison includes the power level, high-gain transmission antenna, two types of antenna diversity implementations, and two different types of DSRC receivers.

The results show that antenna diversity affects the DSRC communication range performance in the blind intersection where the test was conducted. The maximal-ratio combining diversity performs 40-70% better than the packet-switching method in the DSRC communication range measurement. Given a power level of 20 dBm as the nominal transmission power, with maximal-ratio combining diversity implementation, combined with 9 dBi gain antenna, the DSRC communication range can be extended from 95 m to 150 m.

More importantly, this analysis provides additional knowledge and confidence in using DSRC communication in certain intersection collision scenarios. With more than 100 m range in a typical residential 4-way intersection where the average driving speed is 25-30 mph, safety applications could detect and prepare for a warning with 3 to 4 seconds of lead time. This study also shows that as the vehicles moved closer to each other, the transmission power could be increased to provide more robust and longer-range communication thereby further improving the blind-intersection collision warning application performance.

The blind intersection selected in this study is representative as a 4-way residential intersection and the positive outcome of this study is important for further investigation in the two areas. First, more representative intersections, such as urban areas, need to be tested and examined. Buildings surrounding the intersection also influence the reception in the side streets. In some cases, lack of buildings in certain direction of the intersection can lead to missing reflection surfaces and results in considerably less signal power into the crossing street. Second, the high transmission power can cause radio interference. Though interference may not be as critical in a low-traffic residential area, in urban intersections where multiple lanes are present in each direction of traffic, the signal competition levels can be problematic.

The validity of the result is to some extent tied to the data basis. As the DSRC radio becomes more developed, there may be a greater variety of radio chips designed specifically for a mobile environment with more sophisticated antenna diversity implementation.

Future work could involve a diverse selection of representative intersections with various DSRC radio and antenna selections. The measurement of radio interference based on the transmission power could also be a good topic to investigate.

REFERENCES

1. IEEE Computer Society, "*IEEE Standard for Information technology-Telecommunications and information exchange between systems-Local and metropolitan area networks-Specific requirements Part 11: Wireless LAN Medium Access Control (MAC) and Physical Layer (PHY) Specifications Amendment 6: Wireless Access in Vehicular Environments*," IEEE Std. 802.11p- 2010, doi: 10.1109/IEEESTD.2010.5514475.
2. IEEE Computer Society, "*IEEE Standard for Information technology-Telecommunications and information exchange between systems-Local and metropolitan area networks-Specific requirements Part 11: Wireless LAN Medium Access Control (MAC) and Physical Layer (PHY) Specifications*" IEEE Std. 802.11- 2007, doi: 10.1109/IEEESTD.2007.373646.
3. IEEE, "*Draft Standard for Wireless Access in Vehicular Environments (WAVE) - Networking Services*," IEEE Std. 1609.3- 2010
4. IEEE, "*IEEE Trial-Use Standard for Wireless Access in Vehicular Environments (WAVE) - Multi-Channel Operation*," IEEE Std. 1609.4-2006, 10.1109/IEEESTD.2006.254109
5. IEEE, "*IEEE Trial-Use Standard for Wireless Access in Vehicular Environments - Security Services for Applications and Management Messages*," IEEE Std. 1609.2- 2006, doi: 10.1109/IEEESTD.2006.243731.
6. SAE International Surface Vehicle Standard, "Dedicated Short Range Communications (DSRC) Message Set Dictionary," SAE Standard J2735, Rev. Nov. 2009.
7. Alexander, P.; Haley, D.; Grant, A.; "Cooperative Intelligent Transport Systems: 5.9-GHz Field Trials," *Proceedings of the IEEE*, vol.99, no.7, pp.1213-1235, July 2011, doi: 10.1109/JPROC.2011.2105230.
8. Hong, Kezhu, Xing, Daniel, Rai, Vinuth, Kenney, John, "Characterization of DSRC performance as a function of transmit power," *Proceedings of the sixth ACM international workshop on VehiculAr InterNETworking*, September 25-25, 2009, Beijing, China, doi: 10.1145/1614269.1614281.
9. Bai, Fan; Krishnan, H.; "Reliability Analysis of DSRC Wireless Communication for Vehicle Safety Applications," *Intelligent Transportation Systems Conference*, 2006. ITSC '06. IEEE, vol., no., pp. 355-362, 17-20 Sept. 2006, doi: 10.1109/ITSC.2006.1706767.
10. Miucic, R. and Schaffnit, T., "Communication in Future Vehicle Cooperative Safety Systems: 5.9 GHz DSRC Non-Line-of- Sight Field Testing," *SAE Int. J. Passeng. Cars - Electron. Electr. Syst.* 2(1):56-61, 2009, doi:10.4271/2009-01-0163.
11. Mangel, Thomas, Michl, Matthias, Klemp, Oliver, Hartenstein, Hannes, "Real-World Measurements of Non-Line-Of-Sight Reception Quality for 5.9GHz IEEE 802.11p at Intersections", Communication Technologies for Vehicles Third International Workshop, Nets4Cars/Nets4Trains 2011, Oberpfaffenhofen, Germany, March 23-24, 2011
12. Mangel, Thomas, Schweizer, Friedrich, Kosch, Timo, Hartenstein, Hannes, "Vehicular Safety communication at Intersections: Buildings, Non-Line-Of-Sight and Representative Scenarios", 2011 Eighth International Conference on Wireless On-Demand Network Systems and Services, IEEE, 978-1-61284-188

SAE International™

Reliability and Safety/Integrity Analysis for Vehicle-to-Vehicle Wireless Communication

2011-01-1045
Published
04/12/2011

Arkadeb Ghosal, Fan Bai, Rami Debouk and Haibo Zeng
General Motors Company

doi:10.4271/2011-01-1045

ABSTRACT

Vehicle-to-vehicle (V2V) and vehicle-to-infrastructure (V2I) communications are gaining increasing importance in automotive research and engineering domains. The novel communication scheme is targeted to improve driver safety (e.g., forward collision warnings) and comfort (e.g., routing to avoid congestion, automatic toll collection etc.). Features exploiting these communication schemes are still in the early stages of research and development. However, growing attention to system wide infrastructure - in terms of OEM collaboration on interface standardization, protocol standardization, and government supported road/wireless infrastructure - will lead to popularity of such features in the future. This paper focuses on evaluating reliability and safety/integrity of data communicated over the wireless channels for early design verification. Analysis of a design can be done based on formal models, simulation, emulation, and testing. The first two are preferred choices over the later two for early verification, as both formal models and simulation provide rapid verification with high repeatability, fast turn-around time, high control, and high flexibility. Though formal models and simulation may provide lower realism, they help in narrowing down alternatives by discarding outliers. In this paper, we focus on three analysis techniques for early verification/analysis of wireless channels: formal state transition modeling, simulation of the protocol, and hazard and risk assessment. The first two techniques have been considered in several earlier works - in particular, the paper focuses on Markov Chain modeling of the protocol behavior, and simulation of the protocol in ns2. We use both techniques for analyzing a representative vehicle-to-vehicle communication system, and present a comparison of the techniques from the outcome of the analysis. The third technique, hazard and risk assessment (H&RA) approach based on ISO/DIS 26262 has attracted a lot of attention for automotive safety/integrity analysis - however, to the best of our knowledge, the approach has not been widely used with perspective to V2V communication. By applying the H&RA approach to wireless systems, hazards are identified, controls and mitigations are proposed, and safety goals are determined. The controls and mitigations as well as the safety goals will provide requirements on the reliability of the wireless channels and on the communication network architecture.

INTRODUCTION

Vehicle-to-vehicle (V2V) and vehicle-to-infrastructure (V2I) communications are gaining increasing importance in automotive industry. Communication will be based on the IEEE 802.11p standard ([30, 31]), an approved amendment to the IEEE 802.11 standard to add wireless access in vehicular platforms. The wireless communication, also known as Dedicated Short Range Communications (DSRC), is performed over a dedicated channel; in the US, 75 MHz at 5.9 GHz has been set for the purpose. The focus of DSRC is to have connected cars to assist in driving and to increase safety. Figure 1 shows vehicles moving on road while communicating with each other - the shared information may include vehicle conditions (e.g., velocity), road conditions (e.g., slope), and traffic conditions (e.g., accident).

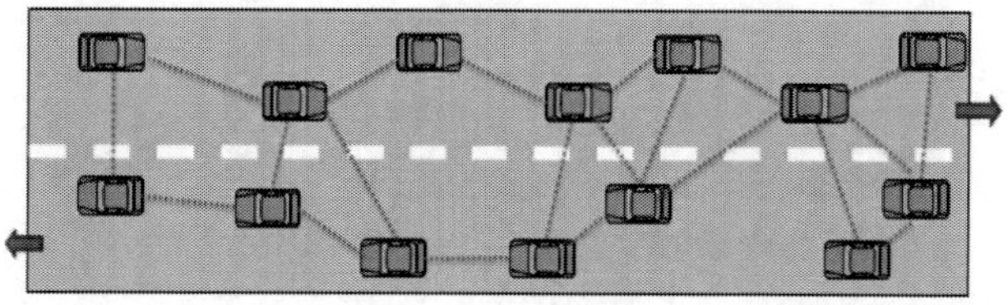

Figure 1. Vehicle-to-Vehicle (V2V) Communication

The communication is not limited to vehicle-to-vehicle information; the protocol can be used for vehicle-to-infrastructure communication schemes like vehicle-to-utilities (e.g., automatic toll payment), vehicle-to-road (e.g., warning on traffic light), vehicle-to-maintenance (e.g., searching for gas stations), vehicle-to-owners (e.g., diagnostic update via cell phone), or vehicle-to-commerce (e.g., targeted local advertisement) (Figure 2).

Figure 2. Vehicle-to-Infrastructure (V2I) Communication

DSRC communication is being targeted for safety applications (e.g., post crash notification, cooperative collision warning, etc), convenience applications (e.g., congested road notification, parking availability notification, etc), and commercial applications (e.g., remote vehicle personalization, diagnostics, etc). Refer [1] for detailed discussion on such possible applications. The applications may also require different types of communication - e.g., unicast vs. broadcast, event-driven vs. periodic, download vs. streaming etc. In summary, the applications (e.g., "360 Degree" Wireless Collision Warning System) are expected to provide a "sixth sense" to the driver (Figure 3) for better and safer driving experience.

Figure 3. "360 Degree" Wireless Collision Warning System

Given the increasing importance of V2V/V2I communication, a holistic framework for performance analysis of wireless channels is of prime importance. Designers need methods and tools for objective evaluation and comparison of automotive ECS (Electronic Control System) architectures to select the most suitable alternative. [2] identifies a set of key top level metrics to compare ECS architectures; the metrics evaluate an architecture based on hard real-time behavior (timing, safety, reliability, etc), ability to accept change (e.g., flexibility, scalability, etc), ability to reflect consumer demands (e.g., fuel efficiency, maintainability, etc), compatibility to legacy designs (e.g., alignment, complexity, etc.), and monetary cost. In this paper, we focus on reliability, and safety-integrity analysis of data communicated over DSRC channels. The objective of the paper is to reflect on the common approaches to compute reliability/safety/integrity of data communicated over DSRC channel, to use the techniques to analyze representative system design, and to compare/evaluate/ analyze outcomes of such analyses.

For reliability analysis, this paper focuses on the metric *packet delivery ratio* (PDR) which measures the fraction of packets that have been successfully received over the channel to the total number of packets that have been transmitted. There are several conventional techniques that are available for computing/assessing the PDR. The techniques can be broadly classified as: formal modeling, simulation, emulation, and test bed implementation. The formal modeling technique usually models the state transition behavior, and captures the general behavior of the state transition systems. The simulation technique records the change in the systems parameters as different events occur - the events are generated randomly to reflect environment behavior. The emulation technique [28, 29] tries to model the system components in software and hardware, and execute the system to analyze the behavior. The test bed [27] implementation technique implements the actual hardware in the real environment, and analyzes traces to estimate patters. The first two techniques are high-level which means they are repeatable, flexible and scalable but may not reflect reality very accurately. The fourth technique reflects reality accurately but is not repeatable, flexible, and scalable. The third technique is in the middle. In this work, we focus on techniques that can be applied to early design verification, when detailed hardware model (either as hardware prototype or software stack required by emulation), or actual hardware (required by test bed implementation) are not available. However given that both formal analysis and simulation are high level, the risk of using any single technique may be high - given that the risk lies in the model, comparing the results across two techniques will provide greater confidence in the results. For our work, we selected two widely used approaches for each technique - Markov Chain modeling for formal analysis, and ns-2 based simulation. Comparing the results of these two approaches, we evaluate how the two models can be used in tandem for early design verification for wireless communication.

Given the systems would be used for safety application, safety analysis need to be performed. In automotive industry ISO/DIS 26262 is the rising standard for such analysis. However to the best of our knowledge there is no open publication on the use of ISO/DIS 26262 process for analyzing wireless applications. In this work, we used hazard and risk assessment (as prescribed by ISO/DIS 26262), and analyzed potential use to wireless application. The paper presents the analysis and the outcome of the analysis for representative wireless applications.

Overview: The next section presents the DSRC protocol. The following two sections provide the aspects of simulation, and Markov Chain modeling analysis for DSRC with reference to related work. The section "Analysis Comparison" provides the outcome of the two analysis techniques to representative data sets, and comparison between the two techniques. The following section presents safety-integrity analysis for wireless data based on ISO/DIS 26262 hazard and risk assessment.

PROTOCOL

In this section, we discuss a brief overview of the protocol; refer [30, 31] for detailed discussion on unicast and broadcast protocols. If a message is ready to be transmitted, then the channel is checked. If the channel is sensed busy, then the sender needs to perform a backoff procedure. The node computes the backoff window length (the length of the window is chosen randomly from an initial allowable window size *Cwin*) and the node waits for a Distributed Inter Frame Space (DIFS) interval. The backoff counter is started at the end of the DIFS. The node monitors the channel for discrete time interval known as slot time (set by the protocol). If there is channel activity within a slot, then the backoff counter process is suspended - once the medium is idle for a DIFS period, the backoff procedure resumes. If there is no activity during the slot interval, then the backoff counter is decremented. When backoff counter reaches zero, the message is transmitted. If the channel is not busy, then the node monitors the channel for DIFS interval. If the channel becomes busy within (or at the end) the DIFS interval, then the node performs a backoff. If the channel is not busy within and at the end of the DIFS, then the node transmits the message. After transmission a new backoff is performed even if there is no other frame waiting to be sent to prevent channel capture.

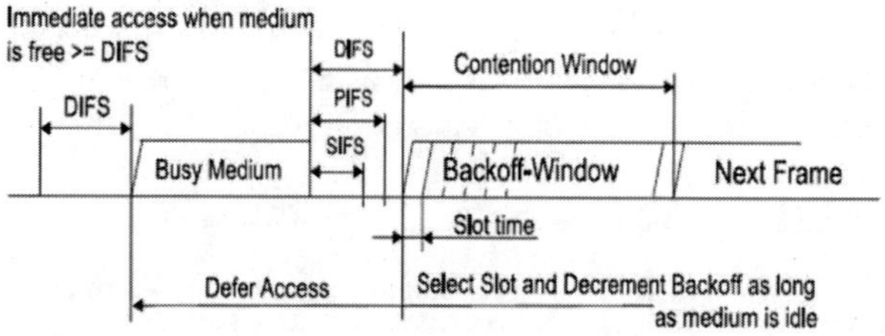

Figure 4. Illustration of Basic Access Method for Scheduling Packet Transmissions (© 1999 IEEE, reprinted from IEEE 802.11 master document, 1999 version [31])

The throughput or reception rate for broadcast protocol may suffer from the following:

• Hidden Terminal- a receiving vehicle may be within range of two transmitting vehicles both of which are outside the transmit range of each other; this can cause packet collision and lost data (which impact throughput).

• Consecutive Freeze Process - under saturated condition, a transmit node may "hog" the channel by selecting zero backoff window.

Note that both the above situations exist in unicast and broadcast protocols; however the lack of acknowledgement makes the situations worse for the broadcast scheme. The automotive environment can bring in other challenges like

• Mobility - the vehicles are almost continuously moving which may mean that they are moving from range of one vehicle to another which will affect packet drop rate;

• Channel Fading - given the traffic, infrastructure, and road condition, ideal channel scenarios would not hold, and it gets difficult to efficiently model channel fading which is needed to compute accurate packet drop ratio;

• Number of cars and distribution - higher the car density, higher would be the number of vehicles trying to transmit at the same time causing higher number of packet collisions;

• Number and type (priority access, arrival rates etc) of messages - depending on origin of signals, messages will be created with varying rates; also, depending on urgency and importance, messages will be assigned different priorities which will detect different access mechanism to MAC; these factors need to be accounted for accurate modeling of delay and throughput. Depending on data arrival rate, the channel can either work under saturated condition (data is always ready to be transmitted at each node), and unsaturated condition (data may not be always available at nodes).

Any analysis for the DSRC broadcast protocol needs to account the above. For a formal model, one needs to account for each of the perspectives. For simulation, one need to account for all type of events such that the above scenarios can occur with realistic probability; the simulation engine should also be capable of handling many different driving condition and transmission loads.

SIMULATION

We implement a generative, parameterized freeway mobility model in a customized Monte-Carlo mobility generator to create synthetic mobility traces. These traces are used to drive simulation study presented in this section. In the generator, vehicle movement is simulated on a straight freeway segment with a stretch of 4 km, where each direction (east-bound or west-bound) has 4 lanes. On each direction, vehicles follow an exponential distribution with exponent λ. The exact lane on which vehicles are placed is randomly determined. We also assume that the vehicle speed follows a Gaussian distribution $N(v, \sigma)$, where v is the average vehicle speed on this direction and $\sigma = v \times 10\%$. Vehicles exiting from one end of the freeway automatically enter the other end. In our study, each run of simulation lasts for 300 second. To remove any random effect, we used different random seeds to generate five vehicle mobility traces for the same set of λ and v values.

Using synthetic mobility traces, we evaluate the impact of vehicle density via ns-2 [26] simulation (version 2.33, with the recently overhauled IEEE 802.11 simulation engine [24]). We also modified several detailed implementations to represent DSRC in accordance with the IEEE 802.11p (WAVE) [22] and IEEE 1609.x (DSRC) draft standard [23], as follows: the carrier frequency is set to 5.9 GHz and the channel bandwidth is halved to 10 MHz. All communication was fixed at 6 Mbps (QPSK modulation scheme - 1/2 rate convolutional coding) with a receiver sensitivity of −93 dBm. A benign Rician fading channel model is used in the simulations. The simulation is performed under two different transmission power scenarios: in one scenario, the transmission power is set to 10 dBm, resulting in an effective transmission range R of 175 meters; in the other scenario, the transmission power is set to 20, resulting in an effective transmission range of R of 500 meters. Two different periodic packet broadcast rates are studied: one with an inter-arrival rate of 100 ms (i.e., 10 packets/sec), and the other with an inter-arrival rate of 500 ms (i.e., 2 packets/sec). The payload of packets (without headers across layers) is set to 214 bytes, in accordance with industry practice [25].

ANALYTICAL MODELS

The seminal work in this area is based on Markov models [3] of unicast protocol assuming saturated condition and idealized channel behavior. The results presented in [3] have inspired many other associated techniques. Some of the later methods used other models like probabilities [4], data-rate switching [5], etc., but Markov model remained the most used. The work has been extended to include lossy channel [6] and freezing [7, 8] under saturated condition. The model has been extended for non-saturated condition in [9, 10] which has been later extended for hidden nodes [11, 12]. The main issue that arises from non-saturated condition is to include a reasonable buffer model. In that effect [13] discusses different aspect of buffer modeling.

A similar modeling has been done for broadcast in [14, 15, 16, 17, 18, 19, 20]. The initial work dealt with separating the saturation [14] and non-saturation model [15], and their effects on analysis for broadcast protocol. However, in the later papers the models account for hidden terminals, channel fading and mobility. In particular, [19] accounts for modeling under saturated condition accounting for hidden terminal and channel fading, and [20] accounts for non-saturated condition accounting for hidden terminal, channel fading and mobility. To the best of our knowledge [19] and [20] are the latest works discussing analytical models for DSRC broadcast protocol. We will use the analytical models discussed in [19] (equation 15) and in [20] (equation 37) for our case study.

ANALYSIS COMPARISON

We have evaluated and compared the results under three different broadcast scenarios:

- Scenario 1: Transmission range = 500 m, and packet arrival rate = 2 packets/s
- Scenario 2: Transmission range = 500 m, and packet arrival rate = 10 packets/s
- Scenario 3: Transmission range = 175 m, and packet arrival rate = 10 packets/s

The transmission range is derived from the power of the transmission. The simulation analysis computes PDR over a range of distances (say from 0 m to 1000 m with a resolution of 25 m); however the analytical model computes a single PDR denoting the average delivery rate for the given transmission range. For the comparison, the simulated data is converted to a single number at a range by computing the average of the PDR over the range. For example, for the first scenario, the simulation produced results at 0, 25, 50, 100,…, 425, 450, 475, and 500 m; a curve fitting method was used to compute the average of the data over that range.

The analytical models are based on the papers [19] and [20]. [19] provides a model assuming saturated data arrival, minimal consecutive freeze effect, and presence of hidden terminals; however the modeling ignored mobility and channel fading. In [20] the authors provide a model for unsaturated data arrival accounting for hidden terminals, channel fading, and vehicle mobility. In [20],

- the channel fading is modeled assuming that a fixed bit error rate is available; however for our simulation setup, the fixed bit error rate model was not used,

• the mobility effect is modeled assuming average relative velocity of two vehicles in the network is constant; however for our simulation setup, the constant average relative velocity could not be validated.

In the analytical model computation, we only used the effect of hidden terminals: i.e. using Equation 15 from [19], and using part of Equation 37 in [20] ignoring channel fading and mobility. Of course this would give rise to a difference between the simulated data and the analytical data. To close the gap, we used an approximation factor *AF* calculated as the ratio of the result computed by analytical model and the simulated data at lowest available car density (which was . 00225 vehicle/ meters in our case), and multiplied the analytical result with the approximation factor *AF*. The intuition behind *AF* is that at very low vehicle density, the effect of channel fading and mobility would be the highest, i.e., the PDR would be dominated by the above factors than packet collision from simultaneous transmission from vehicles competing for channel access. In all of the graphs shown below, the blue line denotes the simulation data, the red line shows the data computed from the analytical model (ignoring channel fading and mobility), and the green line shows the effect of introducing the factor *AF* on the analytical data. To analyze the correlation of the data, we used Pearson correlation (Table 1).

Scenario 1

Figure 5 (resp. Figure 6) shows the comparison of the results for scenario 1 based on the analytical modeling assuming unsaturated (resp. saturated) packet arrival. For this scenario, the arrival rate of the packets is low (2 packets/sec), and the transmission range is 500 m. When the unsaturated model is used, the simulation data and the analytical data have very high correlation (.997807). In fact if the analytical data (red line) is recalculated with *AF*, then the new data (green line) matches with the simulation data (blue line in Figure 5) very closely. When the saturated model is used there is a significant gap with the simulation data - the correlation factor is .923856. Note that while the simulation data is relatively flat, the analytical model (assuming saturation) evaluates a significant drop in PDR. Clearly for this case, the analytical model assuming non-saturation matches with the studied simulation data.

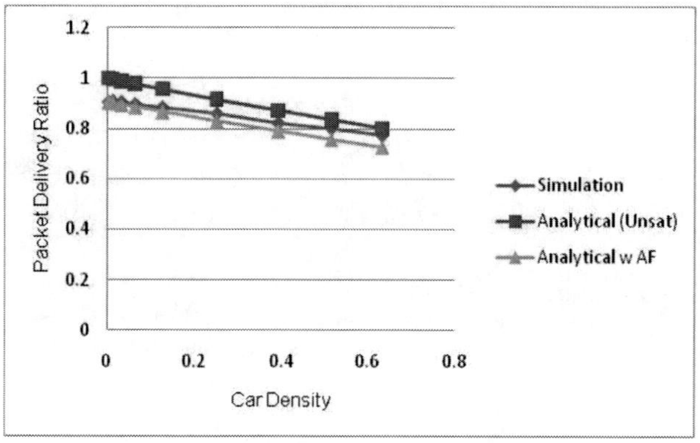

Figure 5. Comparing simulation and analytical (unsaturated model) results for scenario 1

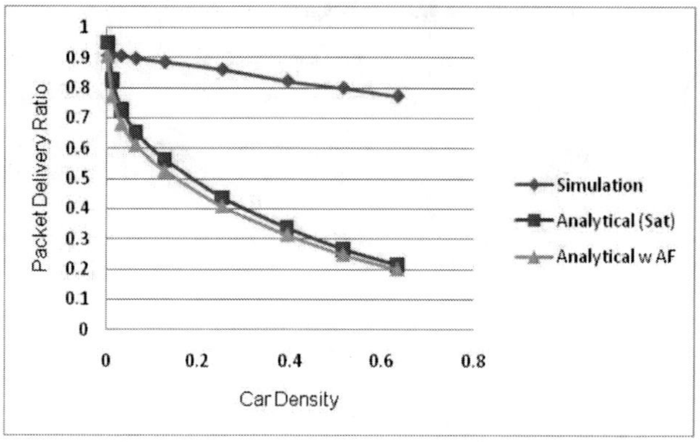

Figure 6. Comparing simulation and analytical (saturated model) results for scenario 1

Scenario 2

Figure 7 (resp. Figure 8) shows the comparison of the results for scenario 2 based on the analytical modeling assuming unsaturated (resp. saturated) packet arrival. For this scenario, the arrival rate of the packets is relatively high (10 packets/ sec), and the transmission range is 500 m. When the unsaturated model is used, the simulation data and the analytical data still has high correlation (.992836); however the correlation is less than what we observed for scenario 1. In fact if the analytical data is recalculated (with the approximation factor), then the data shows similar trends with the simulation data (blue and green line in Figure 7). When the saturated model is used there is a gap with the simulation data - the correlation factor is .945328; however there is better match than what we notice in scenario 1. When we study Figure 8, the curves show similarity in trends at higher car density. This comparison shows that with higher packet arrival rate at high car density, the saturation model is getting similar to the simulation data. In this scenario, the unsaturated model closely tracks the trend of simulation data for low car density, while the saturation model closely tracks the trend of simulation data for high car density.

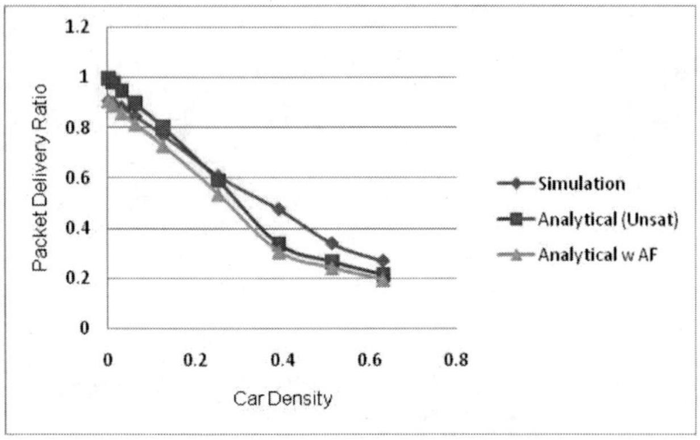

Figure 7. Comparing simulation and analytical (unsaturated model) results for scenario 2

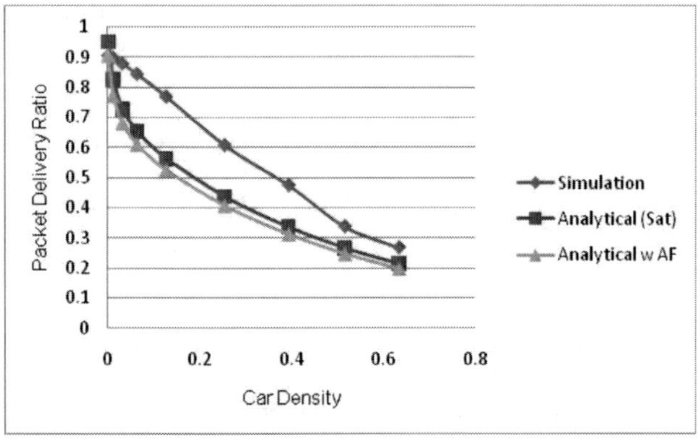

Figure 8. Comparing simulation and analytical (saturated model) results for scenario 2

Scenario 3

Figure 9 (resp. Figure 10) shows the comparison of the results for scenario 3 based on the analytical modeling assuming unsaturated (resp. saturated) packet arrival. For this scenario, the arrival rate of the packets is relatively high (10 packets/sec), but the transmission range is 175 m. When the unsaturated model is used, the simulation data and the analytical data have high correlation (.995131); when the analytical data is approximated with *AF* (green line), the data seems to coincide with the simulated data (blue line). When the saturated model is used, there is gap with the simulation data - the correlation factor is .902750. There is little similarity in trend at low density but there is relatively high similarity in trend at high density. The analytical data is pessimistic as saturated data arrival has been assumed contrary to the actual scenario which is unsaturated. In other words, for this scenario, the unsaturated model works better at low vehicle density, while the saturated model works better at higher vehicle density.

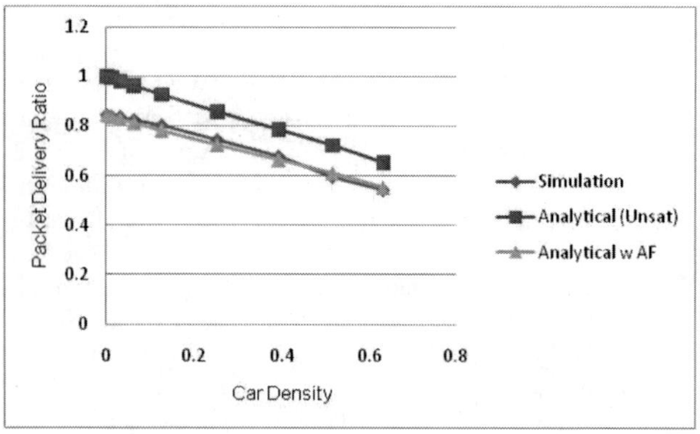

Figure 9. Comparing simulation and analytical (unsaturated model) results for scenario 3

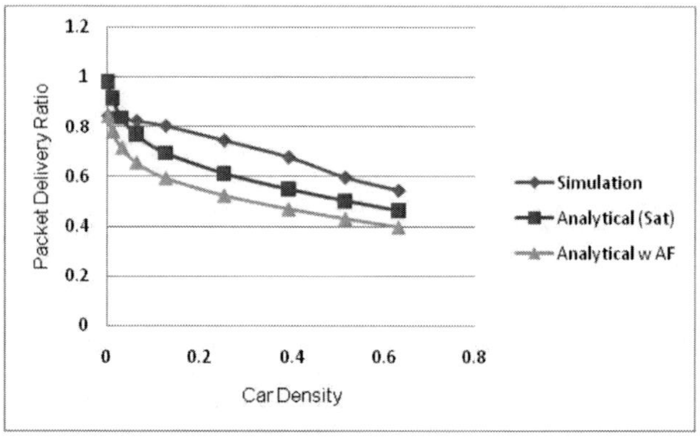

Figure 10. Comparing simulation and analytical (saturated model) results for scenario 3

Table 1. Pearson correlation between simulation and analytical data

Scenario/Packet Arrival Rate	Scenario 1	Scenario 2	Scenario 3
Unsaturation	.997807	. 992836	.995131
Saturation	. 923856	.945328	.902750

The case study shows strong correlation between the analytical model and simulation data. For unsaturated data with sparse arrival rates, there is high degree of similarity between the Markov Chain modeling in [20] and simulation. The simulation was mostly done for unsaturated data, so the saturated model in [19] could not be fully tested but there is strong correlation between the model and the simulation for unsaturated data with high message arrival rates and high vehicle density. The channel fading and mobility models from [20] could not be tested but the approximation helped in showing similar trends between the data computed from the Markov chain modeling and the data gathered from simulation.

SAFETY/INTEGRITY ANALYSIS

As described in [1], DSRC is being targeted to different types of automotive applications ranging from commercial applications to vehicle safety. Commercial applications such as personalization and remote diagnostics are intended to provide driver convenience, and have been in production for a couple of years. However, features related to vehicle safety that not only provide driver convenience but also assist the driver in performing maneuvers, are still being in the research and development phase. These features are in fact safety-critical and hence they must be carefully designed, analyzed, and verified because they are complex, and exhibit new and unique failure modes and effects due to the usage of wireless channels to communicate. To meet the above requirements, a rigorous safety analysis process must be followed.

ISO/DIS 26262, Functional Safety - Road Vehicles [21], is emerging as the standard for functional safety in the automotive electronics domain. Part 3 of [21] describes the process of hazard and risk assessment (H&RA). Potential hazards are identified following an analysis of the operational situations of the system[1]. The system may be a vehicle, a vehicle system, or a vehicle function. For purposes of our analysis, the H&RA process is being applied to a vehicle system or a feature. The identified potential hazards of the feature are then categorized based on the following factors: severity, probability of exposure and controllability. In [21], Severity is defined as the extent of harm to an individual in a specific situation while probability of exposure is defined as being in an operational situation that can be hazardous if coincident with the failure mode under analysis. Controllability is defined in [21] as the avoidance of the specified harm or damage through the timely reactions of the persons involved. The categorization results in the determination of an Automotive Safety Integrity Level (ASIL) to the potential hazard. The ASIL is also assigned to the safety goal(s) formulated to prevent or mitigate the potential hazard, in order to avoid unreasonable risk. Risk reduction (safety) requirements are then derived from these safety goals and inherit their ASIL.

The approach proposed by ISO/DIS 26262 for H&RA, is divided into the following phases:

- Situation analysis and hazard identification
- Hazard classification
- ASIL determination
- Safety goal formulation

The operational situations and operating modes in which the feature may malfunction are to be considered since the malfunctioning behavior may trigger potential hazards. This is achieved irrespective of the communication medium used by the feature. These situations and the corresponding potential hazards are then evaluated. If the feature uses a wireless communication medium then the evaluation process needs to account for that: once a hazard is identified as part of H&RA, causes related to the usage of the wireless channels that may lead to the hazard are to be considered. Some hazard causes when a wireless communication medium is used can be dropped or corrupted messages for instance. Requirements on the reliability and availability of the wireless channels as well as checks on data sanity are derived to control and mitigate these causes. Safety goals to address these aforementioned causes are generated consequently.

As an illustrative example (Table 2), let us assume that the vehicle safety feature is to provide collision mitigation braking through the knowledge exchanged between a cluster of neighboring vehicles. A potential hazard in that situation may be the unintended application of braking. Among other causes leading to the hazard, one may consider a corrupted exchange on the wireless channel between two vehicles. The corrupted message may be due to some electromagnetic interference on the channel. An ASIL for that situation can be determined following the severity, exposure, and controllability categorization of the situation at hand. Controls and mitigation can be proposed such as using parity bits and/or CRCs (Cyclic Redundancy Checks). An example of a safety goal in that case would be to require the messages to be transmitted with enough power. Another potential hazard may be the loss of brake application. Among other causes leading to the hazard, one may consider a dropped message on the wireless channel: the message intended to warn the vehicle about a stationary object on its path is not received, and the collision mitigation braking feature does not apply the brakes. Similar to the discussion of the previous hazard, an ASIL for the new hazard is determined following the severity, exposure, and controllability categorization of the situation at hand. Controls and mitigation can be proposed such as increasing the frequency of retransmission or using the emergency channel. An example of a safety goal in that situation would be to have a channel that prioritizes transmission based on message criticality.

SUMMARY/CONCLUSIONS

In this paper, we studied and compared techniques for early verification of reliability/safety/integrity of data communicated over DSRC channel. We compared analytical models and simulation techniques for analyzing packet

[1] In this paper, the term "system" is used to denote what ISO/DIS 26262 refers to as "item".

Table 2. Hazard Analysis and Mitigation Techniques[2]

Hazard	Scenario	Cause	Controls & Mitigation	S	E	C	ASIL	Safety Goal
Unintended Braking Application	During an exchange of information between vehicles over the wireless channel, spurious radar measurement is transmitted as to mention that the subject vehicle is fast closing on the lead vehicle hence triggering the collision mitigation braking feature	Corrupted wireless message due to EMI (Electro Magnetic Interference)	Wireless channel uses parity bits and CRC checks	X	X	X	X	The wireless channel shall transmit messages with enough power to overcome EMI
Loss of Braking Application	During an exchange of information between vehicles over the wireless channel, a message from the leader vehicle to warn a following vehicle about a stationary object on the road is dropped and hence the collision mitigation braking feature does not receive the input in time to process and apply the brakes to avoid the collision	Dropped wireless message due to limited bandwidth	Use emergency channel for safety-related messages	X	X	X	X	The wireless channel shall provide a priority based transmission scheme

[2]*S = Severity, E = Probability of Exposure, C = Controllability*

delivery ratio. As discussed in the introduction, formal analysis and simulation can be performed during early design phases, and are fast, repeatable, scalable and flexible. However given that both formal analysis and simulation are high level, the risk of using any single technique may be high; given that the risk lies in the model, comparing the results across the two techniques will provide greater confidence in the results. The results show that there is strong correlation between the simulation data and the latest work on Markov Chain based modeling of the behavior of the DSRC broadcast protocol. However, the models have not been tested for all possible assumptions on vehicle and environment behaviors, and one should be careful in using the two techniques in tandem. More studies need to be done (e.g., accounting for correlating different channel fading schemes or effect of vehicle mobility patterns, or using the comparison for other metrics like channel delay and throughput), but the results in this work shows that the overall trends (for both the analyses) are similar.

The paper also discussed how Hazard and Risk Assessment (as dictated by ISO/DIS 26262) can be performed for sub-systems using wireless communication. The discussion on the H&RA technique for a representative example shows the possibility of using ISO/DIS 26262 for wireless subsystems.

REFERENCES

1. Bai, F., Elbatt, T., Holan, G., Krishnan, H., and Sadekar, V.. *Towards characterizing and classifying communication-based automotive applications from a wireless networking perspective*. In Proc. of 1st IEEE Workshop on Automotive Networking and Applications. 2006.

2. Ghosal, A., Giusto, P., Sinha, P., Osella, M. et al., "Metrics for Evaluating Electronic Control System Architecture Alternatives," SAE Technical Paper 2010-01-0453, 2010, doi:10.4271/2010-01-0453.

3. Bianchi, G.. *Performance Analysis of the IEEE 802.11 Distributed Coordination Function.*

4. Zanella, A. and Pellegrini, F. De. *Statistical Characterization of the Service Time in Saturated IEEE 802.11 Networks.* In IEEE Communications Letters. Mar 2005. Vol. 9. No. 3.

5. Xiong, L. and Mao, G.. *Performance Analysis of IEEE 802.11 DCF with Data Rate Switching.* In IEEE Communications Letters. Sep 2007. Vol. 11. No. 9.

6. Dong, X. J. and Varaiya, p.. *Saturation Throughput Analysis of IEEE 802.11 Wireless LANs for a Lossy Channel.* In IEEE Communications Letters. Feb 2005. Vol. 9. No. 2.

7. Ziouva, E. and Antonakopoulos, T.. *CSMA/CA performance under high traffic conditions; throughput and delay analysis.* In Computer Communications. 2002. Vol. 25. pp 313 - 321.

8. Foh, C. H. and Tantra, J. Wirawan. *Comments on IEEE 802.11 Saturation Throughput Analysis with Freezing of Backoff Counters.* In IEEE Communications Letters. Feb 2005. Vol. 9. No. 2.

9. Duffy, K. and Ganesh, A. J.. *Modeling the Impact of Buffering on 802.11.* In IEEE Communications Letters. Feb 2007. Vol. 11. No. 2.

10. Malone, D., Duffy, K., and Leith, D.. *Modeling the 802.11 Distributed Coordination Function in Non-Saturated Heterogeneous Conditions.*

11. Ekici, O. and Yongacoglu, A.. *IEEE 802.11a Throughput Performance with Hidden Nodes.* In IEEE Communications Letters. Jun 2008. Vol. 12. No. 6.

12. Kim, T. and Lim, J.. *Throughput Analysis Considering Coupling Effect in IEEE 802.11 Networks with Hidden Stations.* In IEEE Communications Letters. Mar 2009. Vol. 13. No. 3.

13. Huang, K. D. and Duffy, K. R.. *On a Buffering Hypothesis in 802.11 Analytic Models.* In IEEE Communications Letters. May 2009. Vol. 13. No. 5.

14. Ma, X., Chen, X., and Refai, H.H.. *Unsaturated Performance of IEEE 802.11 Broadcast Service in Vehicle-to-vehicle Networks.* In Proc. of Vehicular Technology Conference. 2007. pp. 1957 - 1961.

15. Ma, X., and Chen, X.. *Saturation Performance of IEEE 802.11 Broadcast Networks.* In IEEE Communications Letters. August 2007. Vol. 11, Issue. 8, pp 686 - 688.

16. Chen, X., Refai, H.H., and Ma, X.. *Saturation performance of IEEE 802.11 Broadcast Scheme in Ad Hoc Wireless LANs.* In Proc. of Vehicular Technology Conference. 2007. pp. 1897 - 1901.

17. Chen, X., Refai, H.H., and Ma, X.. *A Quantitative Approach to Evaluate DSRC Highway Inter-Vehicle Safety Communication.* In Proc. of Global Telecommunications Conference. 2007. pp 151 - 155.

18. Ma, X., and Chen, X.. *Performance Analysis of IEEE 802.11 Broadcast Scheme in Ad Hoc Wireless LANs.* In IEEE Transactions of Vehicular Technology. Vol. 57, Issue. 6, pp 3757 - 3768.

19. Ma, X., Chen, X., and Refai, H.H.. *On the Broadcast Packet Reception Rates in One-Dimensional MANETs.* In Proc. of Global Telecommunications Conference. 2008. pp 1 - 5.

20. Ma, X., Chen, X., and Refai, H.H.. *Performance and Reliability of DSRC Vehicular Safety Communication: A Formal Analysis.* In EURASIP Journal on Wireless Communications and Networking. 2009.

21. *Road Vehicles - Functional Safety*, ISO/DIS 26262 Part 1-10, 2009.

22. IEEE. *Amendment 3: Wireless access in vehicular environments (WAVE).* In Std 802.11 Information Technology Telecommunications And Information Exchange Between Systems, Local and Metropolitan Area Networks. IEEE Press, New York, NY, February 2006.

23. IEEE. *P1609.0-P1609.4 standards (draft).* In *Wireless Access in Vehicular Environments (WAVE).* IEEE Press, New York, NY, 2006.

24. Chen, Q., Schmidt-Eisenlohr, F., Jiang, D., Torrent-Moreno, M., and Delgrossi, L., Hartenstein, H.. *Overhaul of IEEE 802.11 modeling and simulation in ns-2.* In MSWiM 2007. Greece, Aug 2007e.

25. SAE International Surface Vehicle Standard, "Dedicated Short Range Communications (DSRC) Message Set Dictionary," SAE Standard J2735, Rev. Nov. 2009.

26. *The Network Simulator - ns-2.* http://www.isi.edu/nsnam/ns/.

27. *ORBIT.* http://www.orbit-lab.org/.

28. Judd, G. and Steenkiste, P.. *Using Emulation to Understand and Improve Wireless Networks and Applications.* Proceedings of NSDI 2005.

29. Zhou, J., Ji, Z., and Bagrodia, R.. *TWINE: A Hybrid Emulation Testbed for Wireless Networks and Applications.* Proceeding of INFOCOM 2006.

30. ANSI/IEEE Std 802.11, 1999 Edition (R2003) Part 11: *Wireless LAN Medium Access Control (MAC) and Physical Layer (PHY) Specifications.*

31. IEEE Std 802.11p™-2010. Part 11: *Wireless LAN Medium Access Control (MAC) and Physical Layer (PHY) Specifications. Amendment 6: Wireless Access in Vehicular Environments.*

CONTACT INFORMATION

Dr. Haibo Zeng
Senior Researcher
General Motors Company
422 Portage Avenue
Palo Alto CA 94306
Tel: +1-650-207-9383
Fax: +1-650-855-6750
haibo.zeng@gm.com

ACKNOWLEDGMENTS

We would like to thank Dr. Gavin Holland of HRL for his help and suggestions.

DEFINITIONS/ABBREVIATIONS

V2V
Vehicle to Vehicle

V2I
Vehicle to Infrastructure

DSRC
Dedicated Short Range Communication

WAVE
Wireless Access for Vehicular Environments

DIFS
Distributed Inter Frame Space

ASIL
Automotive Safety Integrity Level

H&RA
Hazard and Risk Analysis

PDR
Packet Delivery Ratio

SAE International™

Multi-Sensor System for Vehicle Positioning in Dense Urban Areas

2011-01-1035
Published
04/12/2011

Zeljko Popovic
Honda R&D Americas Inc.

Andrey Soloviev
University of Florida

Yutaka Mochizuki
Honda R&D Americas Inc.

doi:10.4271/2011-01-1035

ABSTRACT

Cooperative vehicle safety can help prevent vehicle collisions by providing timely warnings to the driver or initiating automatic preventive actions based on vehicle dynamics information exchanged between vehicles. The information is shared wirelessly through the emerging DSRC (Dedicated Short Range Communication) standards. The vehicle dynamics information that is shared, such as vehicle velocity and location, is collected from the vehicle's internal sensor communication network and from Global Navigation Systems (GNSS), which includes the Global Positioning System (GPS). GNSS is a critical component of this safety system since it has the needed ability to accurately determine a vehicle's location coordinates in most driving environments. However, its performance can suffer from obstructions in dense urban areas.

Deficiencies of GNSS can be overcome by complimenting GNSS with other sensors. The system presented here combines GPS, Inertial Measurement Unit (IMU), odometer, and laser measurements to achieve lane-level positioning accuracy even in deep urban canyons. This paper introduces the sensor fusion algorithm used to produce the positioning solution, the selection of the optimal sensor suite through simulation, and the performance analysis using test data from a difficult real-world environment.

INTRODUCTION

MOTIVATION: SAVING LIVES

Despite a recent declining trend in total motor vehicle traffic crash fatalities in the United States (from a peak of 43,510 in 2005, which was the highest since 1990, to 37,261 in 2008, which was the lowest since 1961 [1]) and continually declining fatality rates (1.25 fatalities per 100 million miles traveled in 2008, "an all-time low" [1]), the numbers still warrant attention. Some of the short-term decline may be related to economic downturn and it is expected to be cyclical as in the past [1,2]. The long-term declining trend [2] in fatality rate has been attributed to "significant vehicle and occupant safety regulations and programs" and "significant life-saving vehicle technologies like electronic stability control" [1]. The ultimate motivation behind the technologies discussed in this paper is to help accelerate this harm reduction by prevent vehicle collisions.

COOPERATIVE VEHICLE SAFETY

Recent advances in key technologies have allowed demonstrations of vehicle-to-vehicle (V2V) [5] and vehicle-to-infrastructure (V2I) communications-based safety systems [3,4] that enable drivers, through increased awareness of surrounding traffic via warnings, to avoid collisions. More specifically, the idea is that each vehicle continuously wirelessly broadcasts its dynamics in terms of parameters, such as position, velocity, brake status and other information to surrounding vehicles. Each vehicle can assess these parameters to determine if there is a risk of collision. When the risk is significant, the vehicle can warn the driver in visual, audible, or tactile form. In some cases, the vehicle could even automatically attempt to prevent the collision by engaging the brakes, or when a collision is deemed unavoidable, the vehicle could timely prepare safety devices for action, such as pre-tension the seat belts. Such a system could be especially useful when the driver is inattentive or experiences poor driving visibility. Example applications are warnings for those about to pass through a red-light [3,4] and warnings about unseen oncoming traffic when attempting a turn [5]. The key technologies involved in the cooperative collision avoidance approach are standard wireless communication among vehicles and lane-level vehicle positioning for path prediction.

DSRC COMMUNICATIONS

Since cooperative collision avoidance systems are based on sharing operating information between vehicles, a commonly understood wireless communication protocol between vehicles is required. Dedicated Short Range Communications (DSRC) is being established as a communication standard for this purpose. It is low latency wireless communication at the frequency of 5.9 GHz and the range of about 300 to 1000 m depending on environment.

The DSRC physical layer is defined by IEEE standard 802.11p [7] and summarized by others [6]. Standard 802.11p is based on the common WiFi technology [8]. The messages to be transmitted by all enabled vehicles are being defined in the SAE draft standard J2735 [9]. The minimum performance requirements for the transmitted data are being defined in the SAE draft standard J2945, which is in practice use. Operation of DSRC has already been successfully demonstrated in cooperative vehicle safety projects conducted by consortiums of automotive Original Equipment Manufacturers (OEMs) and the U.S. Department of Transportation (US DOT) [3, 4, 5].

GNSS POSITIONING

The critical information that DSRC carries are vehicle positions and velocities. The positioning (including velocity and heading) for cooperative vehicle safety applications needs to be expressed in a common coordinate system, universally available, and sufficiently accurate to distinguish the lane of travel. GPS can meet these needs under certain conditions.

Principles

Of the global navigation satellite systems, GPS is the most widely used and has reliable global coverage [10,11]. GPS is maintained by United States government. The European system, Galileo, is still under development. Other systems, such as Russian GLONASS, still have limitations but may improve in the future.

GPS consists of a set of artificial Earth-orbiting satellites monitored by a network of ground-based stations. There are normally 24 (expanding now to 27 [12]) satellites in GPS implementation, but there could be more or less due to replacement of old satellites. The satellite orbits are precisely (although not exactly) known at all times and broadcast by the satellites themselves. The signal from each satellite also contains a code that allows a user's device (GNSS receiver) to determine the signal's transit time from the satellite to the receiver. The signal travels at the speed of light which allows relating its transit time to the distance to the satellite. Using distances to 3 such satellites, assuming perfect clock synchronization, allows the receiver to determine the user's three-dimensional location. However, for practical reasons, such as cost and size, the receiver clock cannot perfectly maintain synchronization with satellite clocks and thus will have some unknown and varying clock offset. In order to calculate this offset, a measurement of a fourth satellite signal is required. Having at least four measurements provides an estimate of user's three-dimensional position and current time. Using more than four satellites improves this estimate.

In addition to clock biases, there are various other sources of error in determining position using GNSS. The main sources are: uncertainties in satellite positions, the effects of atmospheric conditions on the speed of GNSS signals, receiver noise, and indirect signal paths due to obstructions (termed multipath). Except for receiver noise and multipath, the error sources are common to nearby receivers. This means that differencing positions from nearby receivers can be used to determine the common errors. Although doing this over small distances (several km) provides best results, there are Satellite Based Augmentation Systems (SBAS) that can calculate and disseminate the resulting corrections over a large area (a country) using additional satellites for this purpose. The United States' SBAS service for the GPS system is known as Wide Area Augmentation System (WAAS).

GNSS are complex systems, both in terms of infrastructure involved and the theory behind it. The preceding is a minimal introduction. For a deeper understanding refer to comprehensive textbooks [10-11]. The full specification of GPS-provided signals is given by a U.S. Air Force document [13].

Challenges

The GPS specification guarantees a horizontal error $\leq$ 13 m and vertical error $\leq$ 22 m for 95% of measurements (assuming an unobstructed view of the sky) [14]. This appears far from the accuracy required for vehicle safety needs. However, in practice, receiver manufacturers can claim much better performance. Some examples are 2.0 m CEP (circular error probable) error [15], and 0.6 m RMS (root mean square) error [16]. Independent testing in diverse environments, which mostly excluded dense urban areas, has shown that 1.5 m accuracy is achievable 90% of the time [19]. The 1.5 m accuracy is usually sufficient for cooperative vehicle safety applications because it allows determining the current lane occupied by the vehicle when the lane width is ≥ 3.6 m and driving variation is 0.3 m (Eq. (1)). Fifty-six percent of lanes in the U.S. have the width ≥ 3.6 m (12 ft) [17]. This was calculated from road lengths for all roads that get federal aid, so some local roads are excluded. When the frequency of road use based on road type is taken into account [20], then 1.5 m lane width applies 67% of the time.

$$((1.5 \text{ m error}) + (0.3 \text{ m driving variation})) \times 2 = 3.6 \text{ m} \tag{1}$$

Figure 1. Ten GNSS Receiver Trajectories in Downtown Detroit

However, the availability of useful position estimates can be severely limited where view of the sky is obstructed, such as by buildings or trees. Even when sufficient minimum satellites are visible, such obstructions can still cause multipath errors, which can sometimes be in the order of hundreds of meters. We performed testing of several commonly used receivers in a dense urban environment. The results of this testing have not been published but some trends are worth mentioning to illustrate the magnitude of the problem. The outages (where no position is provided or error is greater than 30 m) for a 2 km test route through a downtown area varied greatly from 0 to 70%, where some receivers commonly had 20 to 40% outages for each test route drive. When position solutions were of sufficient quality to allow across error assessment, the 95% errors for each run were in the order of 10 to 20 m. Figure 1 illustrates the magnitude of the errors by showing 10 sequential trajectories from this testing for a particular receiver.

Urban areas are an important arena for application of cooperative vehicle safety technology due to their high traffic volumes. This is why developing systems that can overcome GNSS deficiencies, such as the one proposed in this paper, are valuable.

PROBLEM STATEMENT

The goal is to achieve lane-level positioning accuracy, even in difficult GPS conditions. There is no standard quantitative expression of lane-level accuracy. Lane widths vary. In the U.S. they can range from 2.7 to more than 3.6 m [17,18]. Error representations for lane-level accuracy encountered in literature and practice range up to 1.5 m [19]. The statistics used for error classification are also diverse, ranging from circular error probable (contains 50% of measurements) to maximum (100% of measurements) error. Error of 2 m for 95% of measurements was taken here as the horizontal aspect of lane-level accuracy. One reason is that this requirement is equivalent to performance abilities for common unaided GNSS receivers in most typical environments [19], and thus it is difficult to expect better in more challenging environments, even with aiding sensors. Another reason is that this accuracy allows a vehicle driven down the center of a 4 m wide lane to be lane-classified correctly most (95%) of the time because being off by 2 m (half of the lane width) from the center of the lane is still within the lane.

The term "difficult GPS environment" is used for its general connotations, but for sake of quantitative assessments, it is defined here as the driving environment where the number of visible satellites is less than 4 or the horizontal dilution of precision (HDOP) is greater than 8. It is also expected that this environment will be severely affected by multipath from the same obstructions that limit the satellite visibility.

Confident use of position data in safety applications also requires knowing estimated bounds on its error. The problem can then be stated as follows: Devise a positioning system that will provide a position estimate in terms of latitude, longitude, height, and quality indication with horizontal error less than 2 m 95% of the time and vertical 95% error that is less than 6 m, even in difficult GPS conditions.

PRIOR WORK

The prior work to address GNSS deficiencies in obstructed environments has involved complementing the GNSS positioning system with one or more sensors that do not depend on open sky conditions. The brief survey that follows introduces some typical approaches and refers to their most recent examples from literature. All these approaches are effective in improving the positioning performance over standalone GPS, however none of them demonstrate quite the sought after accuracy in difficult conditions with an approach that is feasible in automotive domain.

Use of multiple GNSS systems [23,24] can on its own reduce the visibility difficulties in urban environments since having more satellites increases the chance that some will be visible even when only a narrow slice of the sky is open. However, the multipath caused by reflections from buildings can produce significant errors that are not reduced by presence of additional satellites. This is why there are attempts to specifically detect and remove multipath errors [22,34]. Use of receivers that take advantage of increased sensitivity to signals is an approach to gain more satellite visibility [21].

Inertial sensors are the main traditional compliment to GNSS signals for use in navigation. Since inertial sensors only provide relative motion measurements and have errors that are hard to characterize, they need periodic absolute reference updates, which can come from GNSS when it is available. On the other hand, during GNSS outages, the inertial sensors can maintain navigation accuracy for a limited time. Using inertial sensors that are already included on the vehicle [26,28], or of similar quality [25], is an attractive approach. Use of higher-quality sensors extends the tolerable GNSS outage durations [27].

Wheel odometry [28] is often used together with GNSS-inertial combinations. It is convenient because it is usually available on vehicles and is effective because it provides an additional measure of vehicle speed. It is especially useful for detecting when the vehicle is stationary, which can be difficult using GNSS-inertial only.

Use of cameras [29,30] and laser ranging [54,43] are emerging sources of aiding information. Although the information obtained from cameras is less precise than that obtained from laser ranging, cameras are less expensive and are often already available on vehicles.

In addition to GNSS, and in a similar way, it is possible to use other signals for ranging. Some examples are use of DSRC [31,32, 41] and ultra-wide band (UWB) radios [33].

Use of maps has proven effective for general-purpose navigation, and there are continued improvements with their use [35], but safety applications require more accurate maps, which are generally not available commercially.

Beyond aiding sensors, there has been work on using more effective algorithms to improve the positioning output. The most common Kalman filter suffers from its linear assumptions when lower quality sensors are used, so research into non-linear filters [36] has been producing new alternatives. More specifically, a particle filter approach has shown improved results when used with lower-grade sensors [37,38]. Some other recent investigations include use of neural networks [39] and parallel cascade identification (PCI) [40].

The positioning algorithm used here is based on developments already summarily described [43,57]. The theoretical components behind this algorithm are not significantly new. The value of this work resides instead in bringing the theory together into a flexible multi-sensor algorithm that allows practical implementation on a passenger road vehicle and was evaluated in a challenging real-world environment. An earlier paper [42] also describes this implementation, but with different emphasis. More attention is given here to conceptual understanding of sensor use and filtering, as well as the vehicle implementation and the experimental evaluation, while the previous work provides insight into the measurement integrity check and the average velocity calculation from carrier phase measurements.

METHOD AND OUTLINE

The method to devise the required system was to:

1. Assemble an algorithm that can easily accept various combinations of aiding sensors. This algorithm is presented in the section "Multi Sensor Fusion Algorithm".

2. Execute this algorithm with various sensor combinations in a simulated difficult environment in order to determine two best sensor combinations. The five sensor combinations evaluated in this manner were GPS/INS (GI for short), GPS/INS/Odometer (GIO), GPS/Clock/INS/Odometer (GCIO), GPS/Clock/INS/Odometer/Camera (GCIOC) and GPS/Clock/INS/Odometer/Lidar (GCIOL). The two best combinations were found to be and GCIO and GCIOL. This is summarized in the section "Sensor Selection via Simulation".

3. Implement the two best combinations in a vehicle. This is the subject of the section "Vehicle Implementation".

4. Evaluate the solutions in an actual dense urban environment. This is in the section "Experimental Evaluation".

The section "Summary" concludes the paper with an overview of results and potential future steps.

NOTATION

Here are the significant notational conventions that will be followed throughout the following text. The symbol · represents the vector dot product. The *italics* denote a variable. The **bold** face lower case letters are vector variables and bold face upper case letters are matrices. Multiplication is assumed where variables or bracketed expressions are not separated, as in $(ab)(cd)$ = *a* times *b* times *c* times *d*.

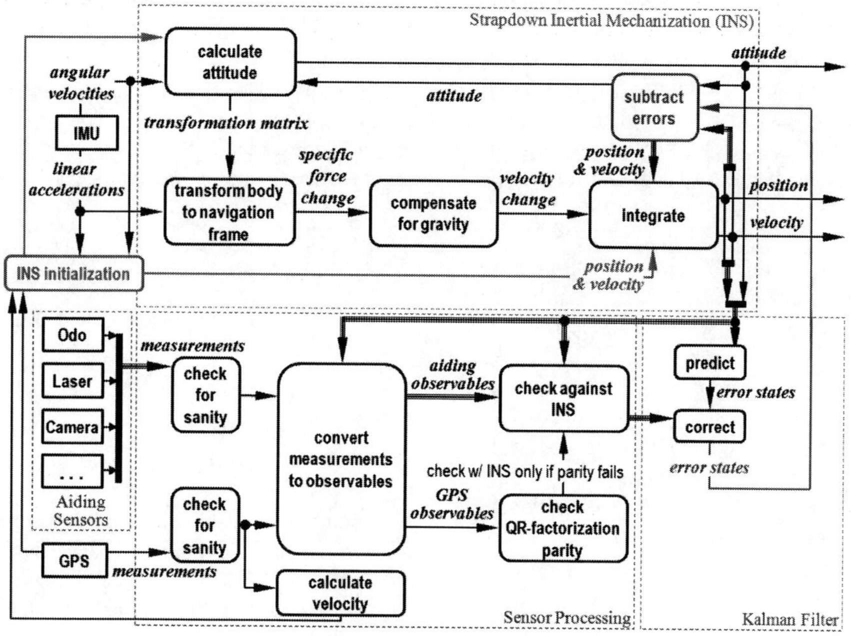

Figure 2. Algorithm's Functional Diagram

MULTI-SENSOR FUSION ALGORITHM

OVERVIEW

The software of the proposed positioning system consists of two components: one component performs communication with sensors to acquire and time-stamp the sensor data (sensor acquisition) while the other component processes this data to derive a position estimate (positioning algorithm). The positioning algorithm is the subject of this section. Sensor acquisition is described in the "Vehicle Implementation" section.

The multifaceted nature of this work prevents exploring fine details of the positioning algorithm within reasonable space, but sufficient attention is given to appreciate the overall architecture and to allow following positioning information from its extraction from various sensor measurements to its formulation into the Kalman filter.

The description in this section provides a functional overview, which is illustrated in Figure 2. Individual algorithm components are described in more detail in succeeding sections.

The proposed system currently uses GPS signals and no other GNSS signals, although it is expandable to use other signals.

The algorithm only requires that inertial measurements are constantly available, while it can deal with appearance and disappearance of measurements from other sensors. Additional sensors can be used to improve the positioning accuracy and availability.

The algorithm is triggered by the arrival of a new set of IMU measurements. The current implementation nominally receives IMU data at 150 Hz, thus the algorithm executes about every 6.67 ms (there is some variability in IMU data transmission). At every execution time-step of the algorithm, the INS computations ("INS" in Figure 2) are applied to estimate the position, velocity, and heading, which are the main outputs of the system. Other outputs indicate the quality of the position estimate (number of satellites visible, HDOP, VDOP, horizontal standard deviation, and vertical standard deviation).

Since inertial measurements only quantify relative motion, INS has to be initialized with an absolute measurement before its outputs can be used (see "INS Initialization" in Figure 2). GPS is used for this purpose. INS is re-initialized when GPS-INS discrepancy exceeds a threshold or when GPS outage duration exceeds a threshold. INS computations, including initialization, are described in the "INS" section.

The INS computations involve correcting the inertial sensor errors. As shown in Figure 2, the parameters for error correction are obtained from the previous output of a Kalman filter [43, 44, 45, 46, 47, 48, 49], which fuses measurements from various sensors, including GPS, odometer, scanning laser, and camera, to estimate the INS error parameters.

The algorithm's architecture in principle allows easy addition or removal of other sensors (see "Aiding Sensors" in Figure

2). It was configured in this study to allow use of standalone GPS (WAAS corrections were used throughout) or GPS receiver with an oven-controlled crystal oscillator (OCXO, or just "clock"), odometer ("odo"), scanning laser LIDAR (Light Detection and Ranging) sensor ("laser"), and video camera data. For each sensor, there is a processing step that converts the acquired sensor data into parameters usable by the Kalman filter.

Checks are performed on all aiding sensors and GPS in order to remove faulty measurements that could corrupt the position estimation. For aiding sensors, this involves comparing an INS-based prediction of the sensor measurement with the actual measurement and then rejecting any measurements that exceed a threshold [42]. In the case of GPS, measurements from individual GPS satellites are checked for consistency using QR-factorization parity [51, 52]. If the parity check fails, INS is used to identify and remove outlying measurements.

KALMAN FILTER

Textbooks that provide a more comprehensive theoretical background on the Kalman filter [48,49] can be used as a supplement to the brief introduction that follows, which was based on other introductory descriptions [10,46,47]. The application of the Kalman filter to the multi-sensor fusion problem has also been explored in the past [43].

Conceptually, a Kalman filter combines estimates of an unknown quantity from different sources by weighing more reliable sources more heavily. The weighing is determined from statistics characterizing the source error. As the basis for the weighing factor, Gaussian distributions are assumed with variance used for single-dimensional cases and covariance for multi-dimensional cases. A simple example of continually estimating the value of a constant x using continually incoming measurements y is shown in the filter of Eq. (2).

$$\hat{x}_{new} = \hat{x}_{current} + K(y_{new} - \hat{x}_{current}) \tag{2}$$

where $\hat{x}_{new}$ is the estimate of x updated using the new measurement, $\hat{x}_{current}$ is estimate established so far, $K = \frac{\sigma^2_{\hat{x}current}}{\sigma^2_{\hat{x}current} - \sigma^2_{\hat{y}new}}$ is the Kalman gain that contains the statistics of the problem, and y_{new} is the newly arrived measurement.

The uncertainty of the measurement is represented as its variance. As the quality of the measurement increases, its variance decreases. The equation shows that as the variance of the measurement decreases relative to the variance of the current estimate, the filter gain weighing the contribution of the new measurement increases and vice-versa. This is the intuitively desired operation. Continuous applications of the equation also show that the filter is recursive, in that all previous measurements affect the current estimate.

In a more realistic example, the goal of the estimation, the state vector $\boldsymbol{x}$, has multiple (n) dimensions (positions, velocities, etc.) that change dynamically. The state is modeled using the linear stochastic difference equation [46]

$$\boldsymbol{x}_k = \boldsymbol{A}\boldsymbol{x}_{k-1} + \boldsymbol{B}\boldsymbol{u}_k + \boldsymbol{w}_{k-1} \tag{3}$$

and corresponding measurement equation

$$\boldsymbol{z}_k = \boldsymbol{H}_k\boldsymbol{x}_k + \boldsymbol{v}_k \tag{4}$$

where x is the $n \times 1$ state vector, $\boldsymbol{A}$ is the $n \times n$ matrix that relates two consecutive states in absence of the control input and noise, $\boldsymbol{u}$ is the optional $l \times 1$ control input (assumed to be zero here); $\boldsymbol{B}$ is the $n \times l$ matrix that relates the control input to the state (also zero here); $\boldsymbol{w}$ is the $n \times 1$ process noise that is white with normal probability distribution, mean of 0 and covariance $\boldsymbol{Q}$; z is the $m \times 1$ measurement vector, $\boldsymbol{H}$ is the $m \times n$ matrix that relates the state to the measurement, and $\boldsymbol{v}$ is the $m \times 1$ measurement noise that is white with normal probability distribution, mean of 0, and covariance $\boldsymbol{R}$.

The estimate error is

$$\boldsymbol{e}_k = \boldsymbol{x}_k - \hat{\boldsymbol{x}}_k \tag{5}$$

and the error covariance matrix $\boldsymbol{P}$, is

$$\boldsymbol{P}_k = \mathrm{E}(\boldsymbol{e}_k\hat{\boldsymbol{e}}_k{}^{\mathrm{T}}) \tag{6}$$

where is E is the expectation function. $\boldsymbol{P}$ has two versions at each step k: one just before the measurement is taken into account (a priori), $\boldsymbol{P}_k^-$, and one after the measurement is included (a posteriori), $\boldsymbol{P}_k$. Statistically, a Kalman filter is optimal because it minimizes the estimated a posteriori error covariance $\boldsymbol{P}_k$ for linear stochastic systems given by Eqs. (3) and (4). A Kalman filter gain that achieves this is:

$$\boldsymbol{K}_k = \left(\boldsymbol{P}_k^-\boldsymbol{H}_k{}^{\mathrm{T}}\right)\left(\boldsymbol{H}_k\boldsymbol{P}_k^-\boldsymbol{H}_k{}^{\mathrm{T}} + \boldsymbol{R}_k\right)^{-1} \tag{7}$$

which is a generalized version of K used in Eq. (2) [46][48].

In general, the application of the Kalman filter consists of two steps. The first step is prediction (or time update), where the estimates of states available at the end of the previous execution are brought to the current time (without including the current measurement) by using a dynamic model of the system. Once INS is initialized, this is done using Eq. (8), which is derived from Eq. (3) with $\boldsymbol{u}$, $\boldsymbol{B}$, and the unknown noise term assumed to be zero:

$$\boldsymbol{x}_k = \boldsymbol{A}\boldsymbol{x}_{k-1} \tag{8}$$

This step also includes updating the estimate error covariance,

$$\boldsymbol{P}_k = \boldsymbol{A}\boldsymbol{P}_k^{-}\boldsymbol{A}^{\mathrm{T}} + \boldsymbol{Q}_k \tag{9}$$

The second step is the correction step (or measurement update, or estimation step) where the new measurement is combined with the results of the prediction step by weighting different values according to their covariances using the Kalman gain of Eq (7) [46]:

$$\hat{\boldsymbol{x}}_k = \hat{\boldsymbol{x}}_k^{-} + \boldsymbol{K}_k(\boldsymbol{z}_k - \boldsymbol{H}_k\hat{\boldsymbol{x}}_k^{-}) \tag{10}$$

$$\boldsymbol{P}_k = (\boldsymbol{I} - \boldsymbol{K}_k\boldsymbol{H}_k)\boldsymbol{P}_k^{-} \tag{11}$$

In this application, if data is available from at least one of the aiding sensors (i.e., GPS, odometer, laser, and camera), then the correction step is performed.

Table 1. Kalman Filter States

Group (States)	Subgroup or State	States
INS Errors (21)	position errors	3
	delta position errors between GPS updates	3
	velocity errors	3
	attitude errors	3
	delta attitude errors between GPS updates	3
	gyro bias errors	3
	accelerometer bias errors	3
GPS Receiver Clock Errors (3)	bias error	1
	drift error	1
	drift error accumulated between GPS updates	1
Odometer Errors (10)	position errors	3
	scale factor	1
	error in orientation to IMU	3
	delta position error between odo and IMU updates	3
Laser Errors 3+(2)(n. of lines)	error in orientation to IMU	3
	line range error	1/line
	lien angular error	1/line

Instead of using the Kalman filter for estimating the states, it can be used to estimate state errors. This is called a complimentary Kalman filter. This approach, applied here, reduces the magnitude of variables and thus reduces numerical errors. It also moves the non-linear navigation problem closer to the assumption of linearity. In this application, the states are sensor errors. The states, listed in Table 1, form the vector state $\boldsymbol{x}$. Their Kalman filter estimate is $\hat{\boldsymbol{x}}$. Components of $\hat{\boldsymbol{x}}$ can be used to subtract out the errors of inertial sensors used in INS and thus improve the positioning solution.

In the complimentary Kalman filter, the values forming the measurement vector z are not actual measurements, but differences between values derived from actual measurements and their predictions based on the INS. This is shown as

$$z = \rho^{INS} - \rho \tag{12}$$

where ρ is the distance-based metric derived from the actual measurement; ρ^{INS} is ρ predicted based on INS, as a function of INS derived position, velocity, and attitude: $\rho^{INS} = f(\boldsymbol{R}_{INS}, \boldsymbol{V}_{INS}, \boldsymbol{\alpha}_{INS})$. To distinguish values forming z from the actual measurements they are referred to as Kalman filter observables. The measurements handled in this way here are GPS, odometer, and laser, all of which can be expressed in terms of distances. The camera measurements, since they do not provide ranges to objects, are handled differently. The derivation of ρ for each of the sensors is described in the sensor sections that follow. Linearization of Eq. (12) can be performed using Taylor series expansion [43].

COORDINATE FRAMES

This section defines the coordinate frames used and the means for converting between different coordinate frame representations.

The vehicle coordinate frame origin coincides with the IMU placement in the vehicle, which is at a central location on the floor inside the vehicle. The *x*-axis is transverse to the vehicle and positive to the right (from the driver side toward the passenger side), the *y*-axis is along the vehicle and positive toward the front, and the *z*-axis is vertical, positive upward. The vehicle coordinate frame is used to express the measurements of vehicle-based sensors. This frame is also referred to as the body frame and noted with *b*. This frame is equivalent to the IMU and the INS frames.

The navigation coordinate frame (*N*) origin is placed at the IMU location at the start of the current test drive. Its *x*-axis is oriented West-East (to East is positive), *y*-axis is South-North (positive), and *z*-axis is down-up (positive). The navigation frame is used to represent the position solution locally, in terms of displacement in meters from the start of the drive.

The Earth coordinate frame (*e*) origin is at the center of the Earth. The *x*-axis is along the line that connects the center of the Earth with the point of intersection between the Greenwich meridian (0° longitude) and the equator (0° latitude) and is positive in that direction. The *y*-axis lies on the line that connects the center of the Earth to the intersection between 90° longitude and 0° latitude and is positive in that direction. The *z*-axis is aligned with the

Earth's polar axis and is positive in the North direction. The Earth frame is used to represent absolute global positions in terms of latitude and longitude.

Given a vector $\boldsymbol{v}_1$ expressed in one coordinate frame, an axis of rotation expressed as a unit vector $\boldsymbol{u}$, and an angle of rotation α, the vector $\boldsymbol{v}_1$ can be expressed as $\boldsymbol{v}_2$ in the second coordinate frame resulting from such rotation, as follows:

$$\boldsymbol{v}_2 = \boldsymbol{C}_1^2 \boldsymbol{v}_1 \tag{13}$$

where

$$\boldsymbol{C}_1^2 = e^{\begin{bmatrix} 0 & -\alpha u_z & \alpha u_y \\ \alpha u_z & 0 & -\alpha u_x \\ -\alpha u_y & \alpha u_x & 0 \end{bmatrix}} \tag{14}$$

is a matrix exponential of the skew-symmetric matrix formed from the components of $\alpha\boldsymbol{u}$. The matrix $\boldsymbol{C}_1^2$ is the direction cosine matrix (DCM) for this transformation.

For two non-collinear vectors, $\boldsymbol{v}_1$ and $\boldsymbol{v}_2$, values of $\boldsymbol{u}$ and α can be found to rotate one vector into the other as follows:

$$u = \boldsymbol{v}_1 \times \boldsymbol{v}_2 \tag{15}$$

$$\alpha = \cos^{-1}\left(\frac{\boldsymbol{v}_1 \cdot \boldsymbol{v}_2}{|\boldsymbol{v}_1||\boldsymbol{v}_2|}\right) \tag{16}$$

Some of the following sections will use DCMs for coordinate frame conversions as in Eq (13).

INS

The IMU provides measurements of linear accelerations along three axis (in m/s^2) and angular velocities along three axis (in °/s) to the INS. The INS, starting from an initial position provided by GPS, integrates these rates over time to estimate current attitude (heading in three dimensions), position, and velocity. This also involves coordinate transformation and gravity compensation. The approach is known as strapdown inertial navigation [50]. "Strapdown" refers to an IMU that is fixed to the frame of the vehicle (as opposed to isolated from vehicle motions).

Initialization

INS first has to be initialized in order to provide an absolute position based on the relative measurements. The following initial parameters are required: position, velocity (speed and direction), and attitude (heading and tilt angles orienting the vehicle with respect to the navigation frame). In this system, GPS is used for position and velocity, after sufficient GPS quality is confirmed via reported HDOP and number of visible satellites for a certain time. The position is used directly as provided by the GPS receiver. The velocity is calculated from carrier phase changes (see the "GPS" section).

The attitude is determined from two non-collinear vectors whose values are known in both the vehicle and navigation frames. One is the velocity vector which is known in the vehicle frame from INS (this is termed $\boldsymbol{v}_{INS}^b$), and in the navigation frame from GPS (this is termed $\boldsymbol{v}_{GPS}^N$). The other is the vector representing position change between the time the last GPS data was received and the current time (the current time is the same as the time of the latest IMU data). This vector is also available in the vehicle frame from INS (labeled $\boldsymbol{R}_{INS}^b$) and in the navigation frame based on GPS velocity and gravity (labeled $\boldsymbol{R}_{GPS}^N$). The attitude is then defined via a DCM matrix,

$$\boldsymbol{C}_b^N = \mathbf{C}_2\mathbf{C}_1 \tag{17}$$

which relates the two frames. First $\mathbf{C}_1$ is found,

$$\boldsymbol{v}_{GPS}^N = \boldsymbol{v}_{INS}^N = \mathbf{C}_1\boldsymbol{v}_{INS}^b \tag{18}$$

using vectors $\boldsymbol{v}_{INS}^b$ and $\boldsymbol{v}_{GPS}^N$ with Eqs. (14), (15), (16) to derive $\mathbf{C}_1$. This also gives $\boldsymbol{R}_{INS}^{б} = \mathbf{C}_1\boldsymbol{R}_{INS}^b$. Then $\mathbf{C}_2$ is found.

$$\boldsymbol{R}_{GPS}^N = \boldsymbol{R}_{INS}^N = \mathbf{C}_2\boldsymbol{R}_{INS}^{б} \tag{19}$$

using $\boldsymbol{v}_{INS}^N$ as the axis of rotation. The angle of rotation is then the angle between the projections of $\boldsymbol{R}_{INS}^{б}$ and $\boldsymbol{R}_{GPS}^N$ onto the plane normal to $\boldsymbol{v}_{INS}^N$. With this axis and angle of rotation, Eq. (13)-(14) can be used to determine $\mathbf{C}_2$.

The attitude DCM resulting from this initialization is updated at each execution time step to reflect the change in attitude resulting from motion since the previous step.

Mechanization

The attitude DCM is updated using $\Delta\boldsymbol{C}$ which is the DCM representing the change in attitude. This DCM is found from IMU angular data as follows. The angular rates ($\boldsymbol{\omega}$ in °/s) from the IMU are converted into angular displacements (°) using the time since the previous IMU data set, Δt, and organized as vector $\boldsymbol{\Delta\Theta}$

$$\Delta\mathbf{\Theta} = \begin{bmatrix} \omega_x \\ \omega_y \\ \omega_z \end{bmatrix} \Delta t \tag{20}$$

These are then corrected for the Earth rate [50]:

$$\mathbf{\Phi} = \Delta\mathbf{\Theta} - (\boldsymbol{C}_b^N)^{\mathrm{T}} \mathbf{\Omega} \Delta t \tag{21}$$

where $(\boldsymbol{C}_b^N)^{\mathrm{T}}$ is used to transform the Earth rotation rate $\mathbf{\Omega}$ over Δt into the vehicle frame. The set of rotation angles $\mathbf{\Phi}$ can then be used to form $\Delta \boldsymbol{C}$, similar to Eq. (14), as

$$\Delta \boldsymbol{C} = e^{\begin{bmatrix} 0 & -\mathbf{\Phi}_z & \mathbf{\Phi}_y \\ \mathbf{\Phi}_z & 0 & -\mathbf{\Phi}_x \\ -\mathbf{\Phi}_y & \mathbf{\Phi}_x & 0 \end{bmatrix}} \tag{22}$$

Then the updated DCM is

$$\boldsymbol{C}_{b\,t}^{N} = \boldsymbol{C}_{b\,t-1}^{N} \Delta \boldsymbol{C} \tag{23}$$

In order to update the velocity, the linear accelerations from IMU ($\boldsymbol{\alpha}$ in m/s^2) are used. They are organized as the change in the specific force vector $\Delta \mathbf{f_b}$ in m/s:

$$\Delta \mathbf{f_b} = \begin{bmatrix} a_x \\ a_y \\ a_z \end{bmatrix} \Delta t \tag{24}$$

The term specific force is used because the effect of gravity has not been removed.

The expression for the change in the DCM over small angles is [50]:

$$\boldsymbol{C}_b^N(\mathrm{t} + \delta\mathrm{t}) = \boldsymbol{C}_b^N(\mathrm{t}) \left(\mathbf{I} + \begin{bmatrix} 0 & -\mathbf{\Phi}_z & \mathbf{\Phi}_y \\ \mathbf{\Phi}_z & 0 & -\mathbf{\Phi}_x \\ -\mathbf{\Phi}_y & \mathbf{\Phi}_x & 0 \end{bmatrix} \right) \tag{25}$$

This equation is used to transform the specific force vector from the vehicle to the navigation frame as follows:

$$\Delta \mathbf{f}_{\mathrm{N}} = \boldsymbol{C}_b^N \left(\mathbf{I} + \frac{1}{2} \begin{bmatrix} 0 & -\mathbf{\Phi}_z & \mathbf{\Phi}_y \\ \mathbf{\Phi}_z & 0 & -\mathbf{\Phi}_x \\ -\mathbf{\Phi}_y & \mathbf{\Phi}_x & 0 \end{bmatrix} \right) \Delta \mathbf{f}_{\mathrm{b}} \tag{26}$$

where 1/2 is inserted to transform to the middle of the time interval as a way of representing average specific force over the interval.

Compensation for acceleration due to gravity ($\mathbf{g}$) is performed using Eq. (27):

$$\Delta \mathbf{V}_{\mathrm{N}} = \Delta \mathbf{f}_{\mathrm{N}} + \mathbf{g}\mathbf{t} \tag{27}$$

Now the change in velocity, $\Delta \mathbf{V}_{\mathrm{N}}$, can be integrated into the current velocity and position estimates, $\mathbf{V}_{\mathrm{t}}$ and $\mathbf{R}_{\mathrm{t}}$, respectively:

$$\mathbf{V}_{\mathrm{t}} = \mathbf{V}_{\mathrm{t}-1} + \Delta \mathbf{V}_{\mathrm{N}} \tag{28}$$

$$\mathbf{R}_{\mathrm{t}} = \mathbf{V}_{\mathrm{t}-1} \Delta t + \frac{1}{2} \Delta \mathbf{V}_{\mathrm{N}} \Delta t \tag{29}$$

These estimates are then used in multiple parts of the algorithm. These include the sensor integrity checks, the formulation of observables, and Kalman filter predictions. They are also outputs of the positioning algorithm (along with their latitude-longitude-height representations and quality measures).

GPS

The data from the GPS receiver that is used in the multi-sensor position calculation includes the GPS-only position solution, satellite positions, and carrier phase measurements.

Pseudorange Measurements

A GPS carrier signal is one of the signals that carries the GPS information. GPS signal used here is a sinusoid L1 at 1575.42 MHz. The intended use of GPS signals is to detect the satellite code superimposed on the carrier signal and then measure the time difference from the start of the code generated by the receiver and the start of the code actually received from a satellite. Since it is known when the satellite and the receiver start code generation, this time difference is mainly due to the signal's transit time that then can be converted to the estimated distance to the satellite, called pseudorange, using the speed of light. Using pseudoranges to multiple satellites allows calculation of the receiver position and clock differences. This is how the GPS receiver used here calculates the position it provides. The effective achievable precision using this approach and accounting for some errors is about 1 m.

Carrier Phase Measurements

Fortunately, there is another measurement provided by the GPS receiver, the carrier phase measurement φ, that is used by the multi-sensor solution proposed here. The carrier phase is the phase difference, expressed as a fraction of the signal cycle, between the carrier signal received from the satellite and reference carrier signal generated by the receiver (Figure

3). The phase can be converted into distance using the signal wavelength. This conversion is very precise as the phase difference can be determined with resolution of 0.01 cycles which is equivalent to 19 mm for the L1 signal. However, phase difference cannot be directly used to estimate the distance to the satellite because it is only possible to directly measure the difference in phase and not also the number of whole cycles in the transit delay. This integer ambiguity (equal to 2 in the example of Figure 3) cannot be determined in real-time for a single receiver, although it is possible to resolve when multiple receivers are used [11].

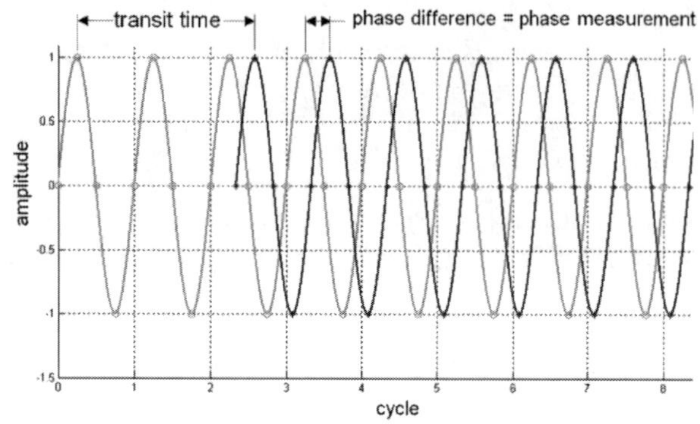

Figure 3. Carrier Phase Measurement

Carrier Phase Kinematics

Fortunately, the change in carrier phase measurement over time, $\Delta\varphi$ (Figure 4), can be used in real-time [43].

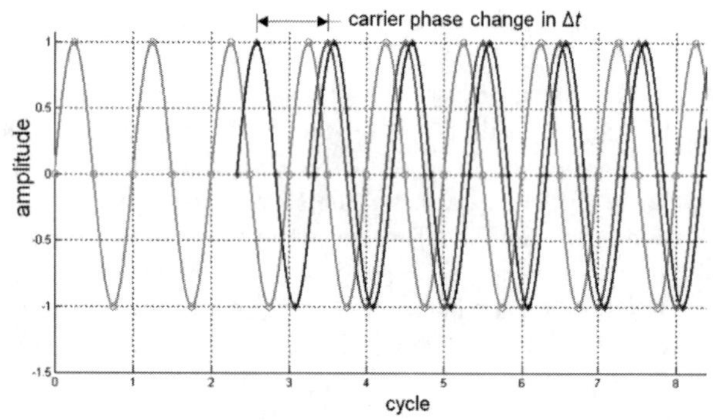

Figure 4. Carrier Phase Measurement Change

Parameter $\Delta\varphi$ represents the change in the relative position between the receiver and the satellite (Δr) but also includes changes in the bias of the receiver clock with respect to the satellite clock ($\Delta\delta t_R$), systematic errors such as atmospheric delays and satellite clock errorrs ($\Delta\varepsilon$), and noise ($\Delta\eta$):

$$\Delta\varphi = \varphi_k - \varphi_{k-1} = \Delta r + \Delta\delta t_R + \Delta\varepsilon + \Delta\eta \tag{30}$$

This equation applies for each satellite. The errors $\Delta\varepsilon$ are either corrected by the GPS receiver (tropospheric errors) or ignored as negligible due to differencing (ionospheric errors). The change in range (Δr) can be expressed in terms satellite motion along the line-of-sight (LOS), Δr_{LOS}, and changes in relative satellite/receiver geometry ($\Delta r_{S/R}$):

$$\Delta r = \Delta r_{LOS} - \Delta r_{S/R} - \boldsymbol{e}_{LOS} \cdot \Delta\boldsymbol{R}_R \tag{31}$$

where $\boldsymbol{e}_{LOS}$ is the unit vector along LOS and $\Delta\boldsymbol{R}_R$ is the change in receiver position based on INS. The terms Δr_{LOS} and $\Delta r_{S/R}$ are calculated as:

$$\Delta r_{LOS} = e_k \cdot \Delta\boldsymbol{R}_{S_k} - e_{k-1} \cdot \Delta\boldsymbol{R}_{S_{k-1}} \tag{32}$$

$$\Delta r_{S/R} = (e_k - e_{k-1}) \cdot \Delta\boldsymbol{R}_{R_{k-1}} \tag{33}$$

where $\Delta\boldsymbol{R}_S$ is the satellite position change based on satellite positions forwarded by the GPS receiver.

Kalman Filter Observables

The parameter used in the formation of the Kalman filter observable is the adjusted phase change. It is defined as:

$$\Delta\varphi^{\text{adj}} = \Delta\varphi - \Delta r_{LOS} + \Delta r_{S/R} \tag{34}$$

Substituting Eq. (31) and Eq. (32) into Eq. (34) also yields:

$$\Delta\varphi^{\text{adj}} = -\boldsymbol{e}_{LOS} \cdot \Delta\boldsymbol{R}_R + \Delta\delta t_R + \Delta\eta \tag{35}$$

This allows estimation of $\Delta\varphi^{\text{adj}}$ using INS as:

$$\Delta\varphi^{\text{INS}} = -\boldsymbol{e}_{LOS} \cdot \Delta\boldsymbol{R}_R^{INS} + \Delta\mathbf{C}_{\text{b}}^{\text{N}}\mathbf{L}_{\text{IMU}}^{\text{GPS}} \tag{36}$$

where $\mathbf{L}_{\text{IMU}}^{\text{GPS}}$ is the vector locating the GPS antenna with respect to the IMU expressed in the vehicle body. Now the Kalman filter observable based on carrier phase change can be formulated using (12) as:

$$z^{CP} = \Delta\varphi^{\text{INS}} - \Delta\varphi^{\text{adj}} \tag{37}$$

A set of z_s^{CP} values over visible satellites s is used as part of the z measurement vector in Kalman filter implementation Eqs. (7), (8), (9), (10), (11).

Note that carrier phase measurements are checked for presence of outliers using a QR-factorization parity check

[51, 52]. If the threshold in the check is exceeded, outliers are assumed to be present and the measurements are then also checked against INS data to identify and remove the outliers [42].

Average Velocity from Carrier Phase for INS Initialization

Carrier phase changes over time are also used here to obtain the average velocity and average clock drift. This velocity is used in INS initialization. Other papers derive the process [51] and summarize the results [42]. Instantaneous velocity is then derived from fitting a polynomial through a sequence of past average velocity estimates.

OSCILLATOR

The GPS receiver clock has bias and drift with respect to GPS time kept by the GPS satellites and ground control stations. These clock error terms are estimated using the Kalman filter. Tightening the statistics of these terms (i.e., reducing their variance) lessens their influence on the position accuracy. The effect may be significant enough to sometimes allow accurate positioning with less than 4 satellites. For this purpose, an oven-controlled crystal oscillator (OCXO) is interfaced with the GPS receiver. The oscillator is expected to have a frequency that is more stable than the one built into the GPS receiver. The provided external oven-controlled 10 MHz oscillator can maintain its reference frequency within $\pm 1 \times 10^{-9}$ Hz over 0 to 50°C and ages $\pm 5 \times 10^{-5}$ Hz over 30 days [53]. The receiver is configured to use the external clock instead of its built-in clock. No additional processing is required in the positioning algorithm (but note that the corresponding value in the covariance matrix is initialized to reflect the improved clock reliability).

ODOMETER

Calculation of a vehicle's wheels' angular displacement is used as an aiding input to the Kalman filter. The main complementing contributions of this measurement is that it provides a confident indication of vehicle at rest (unlike GPS and INS), does not suffer from obstructions (unlike GPS), and does not have hard-to-characterize error (unlike INS). Although it can suffer from the wheel slip condition and inaccurate tire-radius calibration, the slip can be effectively detected and radius continuously calibrated in the Kalman filter.

Kinematics

The odometer is used to keep an independent estimate of vehicle position that is then combined with other estimates in the Kalman filter. This position estimate is initialized using the position obtained at INS initialization. The current odometer count, $count_t$, is read at the rate of 100 Hz. The incremental count, ($count_t - count_{t-1}$), is first converted into incremental distance, Δs, in Eq. (38), by multiplying it with res, the resolution of each count in m.

$$\Delta s = (res)(count_t - count_{t-1}) \tag{38}$$

The conversion of distance into a vehicle displacement vector ($\Delta \boldsymbol{R}$) is derived in the vehicle coordinate frame. It assumes that the odometer measures motion along an arc path in the x-y plane (Figure 5) and the vehicle turns around the z axis on a circle of radius r_{curve}.

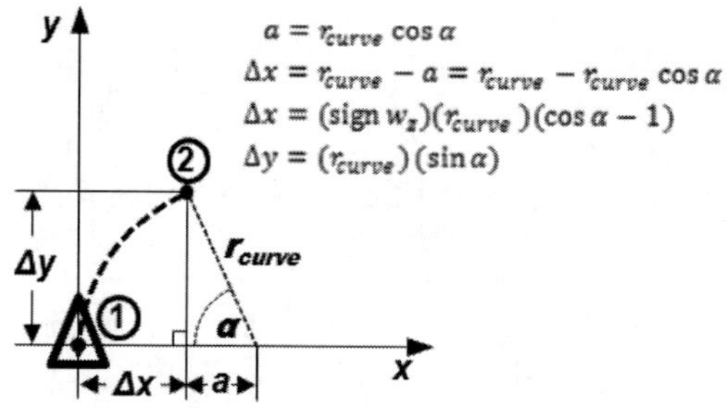

Figure 5. Odometer Measurement Geometry

The resulting Eq. (39) uses vehicle velocity (V) and heading rate (w_z), both from INS:

$$\Delta \boldsymbol{R} = \begin{bmatrix} (\operatorname{sign} w_z)(r_{curve})(\cos\alpha - 1) \\ r_{curve}\sin\alpha \\ 0 \end{bmatrix} \tag{39}$$

where $r_{curve} = \frac{\|V\|}{|w_z|}$ and $\alpha = \frac{\Delta s}{r_{curve}}$.

Then this displacement vector is converted from the odometer into the navigation coordinate frame using Eq. (40).

$$\Delta \boldsymbol{R}_N = \boldsymbol{C}_b^N \boldsymbol{C}_{ODO}^b \Delta \boldsymbol{R} \tag{40}$$

where $\boldsymbol{C}_{ODO}^b$ is the direction cosine matrix for transformation from the odometer frame into the vehicle (body) frame and $\boldsymbol{C}_b^N$ is the direction cosine matrix for the transformation from the body frame into the navigation frame. The new displacement vector, $\Delta \boldsymbol{R}_N$, is turned into odometer-based vehicle position estimate as follows:

$$\boldsymbol{R}_{ODO_t} = \boldsymbol{R}_{ODO_{t-1}} + \Delta \boldsymbol{R}_N \tag{41}$$

Kalman Filter Observable

The Kalman filter observable from the odometer measurements is:

$$z^{ODO} = \boldsymbol{R}_{\text{ODO}}^{\text{INS}} - \boldsymbol{R}_{\text{ODO}} \tag{42}$$

where

$$\boldsymbol{R}_{\text{ODO}}^{\text{INS}} = \boldsymbol{R}_{\text{INS}} + \boldsymbol{C}_b^N \mathbf{L}_{\text{IMU}}^{\text{ODO}}$$

and $\boldsymbol{R}_{\text{ODO}}^{\text{INS}}$ is the INS-based estimate of odometer-based vehicle position, $\boldsymbol{R}_{\text{INS}}$ is INS-based vehicle position expressed in the vehicle body frame, and $\mathbf{L}_{\text{INS}}^{\text{ODO}}$ is the lever arm between the odometer and the IMU expressed in the vehicle body frame. The observable z^{ODO} is part of the measurement vector z, which is used in the filter implementation equations listed in the "Kalman Filter" section.

The Kalman filter takes into account that INS errors influence the odometer-based estimate through the Kalman filter covariance matrix elements that relate INS and odometer measurements.

LASER

Measurement Principle

The laser sends laser light pulses and measures the time it takes for them reflect back to the laser. This time measurement is converted into the desired range measurement by multiplying by the speed of light.

The laser is oriented to scan in an approximately horizontal plane and is positioned on the forward roof of the vehicle. It scans a forward view of 180° at the angular resolution of 1°. The laser data comes as a vector of 181 ranges, where each range is the distance from the laser to the first reflective obstacle it encounters, for each of the 181 angles of the scan (0°, 1°, … 180°). Thus the laser data can be thought of as an array of points in polar coordinates with the origin at the laser.

At each arrival of new laser data, it is first organized (pre-processed) so that it can be used for navigation purposes. Figure 6 shows an overview of this process.

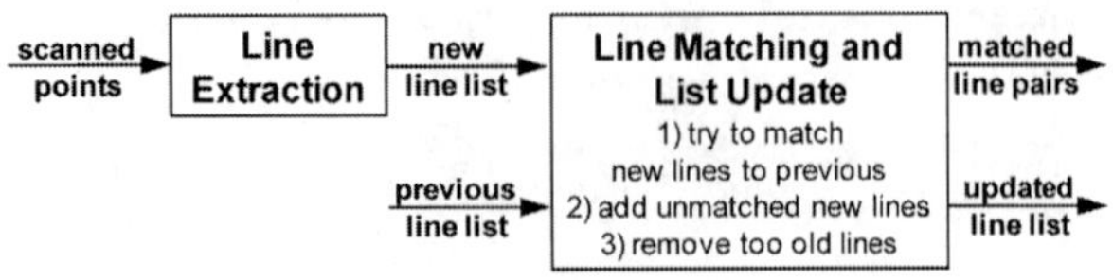

Figure 6. Pre-Processing of Laser Measurements

Line Extraction

Line extraction [54] is the first step. It involves organizing the scanned points into lines as shown in Figure 7.

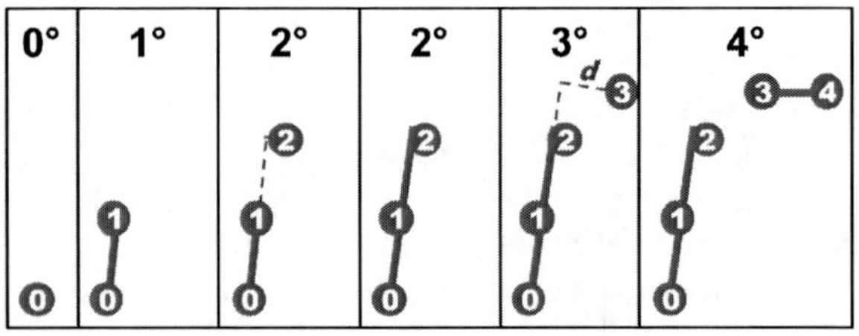

Figure 7. Line Extraction, Illustration of Several Steps

Line extraction starts with the first two scanned points and creates a line between them. The distance (d in Figure 7) of the next point to the line is considered. If it is less than a threshold, it is added to the line (2° in Figure 7) and the line is adjusted using Least Mean Squares [54] to fit the new point. Otherwise, a new line is started (3° and 4° in Figure 7). This is continued until all the points are processed by assigning them to lines. This process creates a list of lines for the current scan where each line is parameterized using its distance from the laser (ρ) and its angle to the laser (α) (Figure 8). The coordinate frame in Figure 8 has its origin at the location of the laser. The figure shows that the distance ρ is the normal to the extracted line that passes though the origin and that the angle α is between the laser x-axis and the normal.

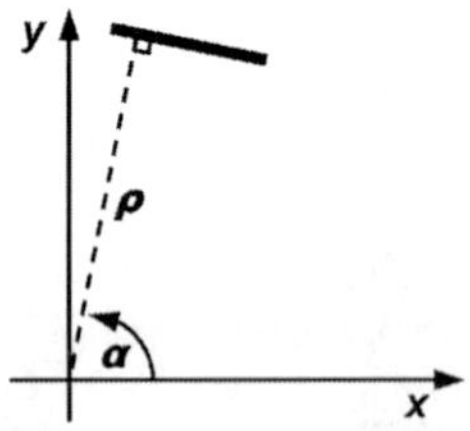

Figure 8. Line Parameters

Figure 9 illustrates line extraction for portion of one actual scan. The scan points are plotted in the middle of the figure looking from above. The photographs on the left and right show the left and right sides of the corresponding actual street scene.

Line Matching

Line extraction is followed by line matching [54]. Line matching first uses INS-based navigation to transform the lines in the line list accumulated from previous scans in order to account for the motion between the previous scan and the new scan. The lines are thus brought to the time of the new scan. Line matching then consists in taking the lines resulting from the new scan and comparing their parameters with the

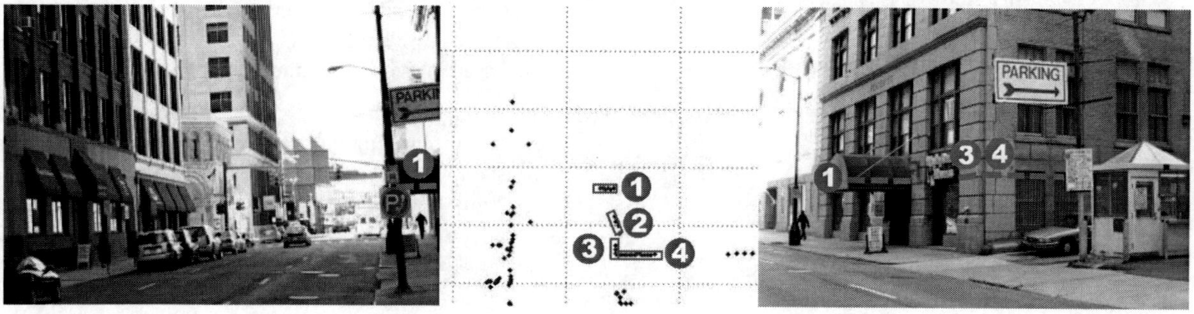

Figure 9. Line Extraction, Real Example

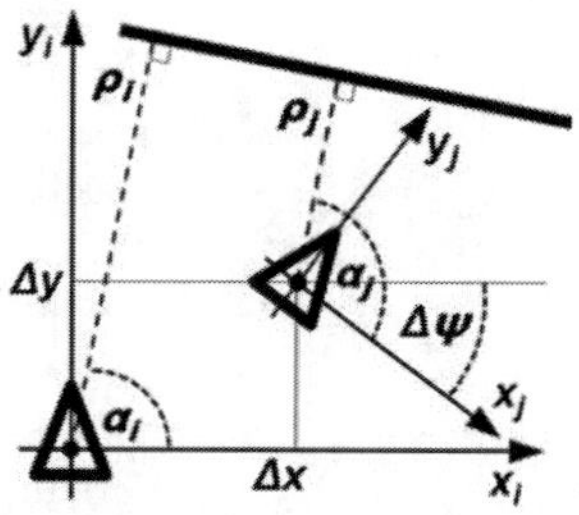

Figure 10. Geometry of Laser Ranging During Motion

parameters of the transformed existing lines. If the parameters are within thresholds for a {(new line), (transformed previous line)}pair, then this is considered a match and the pair of lines {(new line), (original previous line)} is used to derive relative motion measures. That is, the lines in the latter pair are assumed to represent the same line feature on a stationary object but at different moments in time so that any difference in their position is due to the relative motion during that time interval.

Any unmatched new lines are considered to belong to new objects and are transformed to the navigation frame and added to the cumulative line list of previous lines. This line list is pruned for lines than are older than a threshold in order to keep it computationally manageable. The old lines can be thought of as belonging to objects that have likely gone out of view and are not likely to be matched to new lines.

Kinematics

In order to appreciate extracting relative motion from matched lines, consider first the motion of vehicle in Figure 10. The two coordinate systems x_i-y_i and x_j-y_j represent the two-dimensional position and attitude of a vehicle (thick triangle) at two successive laser measurement times i and j. An actual line feature on only one obstacle object (thick line) is considered. At time i, a line is extracted from the laser reflections off this object with parameters (ρ_i, α_i). At time j, a line is matched with parameters (ρ_j, α_j) to belong to the same object using error-affected knowledge of vehicle relative motion from INS. Even though the same object is detected, the line parameters change due to the motion of the vehicle.

The equation of the detected line feature can be expressed in the i frame using the normal form for the equation of a line [55]:

$$x \cos\alpha_i + y \sin\alpha_i = \rho_i \tag{43}$$

where x and y are i frame coordinates of a point on the line. The distance d_0 in frame i from the point (x_0, y_0) to this line expressed in the form $Ax + By + C = 0$ as $(\cos\alpha_i)x + (\sin\alpha_i)y - \rho = 0$ can be determined from Eq. (44) [55].

$$d_0 = \left|\frac{Ax_0 + By_0 + C}{\sqrt{A^2 + B^2}}\right| \tag{44}$$

which yields

$$\begin{aligned} d_0 &= \left|\frac{(\cos\alpha_i)x + (\sin\alpha_i)y - \rho}{\sqrt{(\cos\alpha_i)^2 + (\sin\alpha_i)^2}}\right| = \\ &= |(\cos\alpha_i)x + (\sin\alpha_i)y - \rho| \\ &= \rho - x\cos\alpha_i - y\sin\alpha_i \end{aligned} \tag{45}$$

At time j, the vehicle in Figure 10 is known to be at the i-frame coordinates (Δx, Δy) and to be distance ρ_j from the line. Substituting those parameters in Eq (45) gives

$$\rho_j = \rho_i - \Delta x \cos \alpha_i - \Delta y \sin \alpha_i$$

which is also

$$\rho_i - \rho_j = \Delta x \cos \alpha_i + \Delta y \sin \alpha_i \tag{46}$$

Equation (46) relates the laser range measurements to one stationary object line, at two different times, to x and y components of the motion between those times. It has two unknowns, x and y, and thus at least two line features need to followed from one laser scan to another. That is, at least two matched pairs are needed. Increasing the number of line pairs increases the accuracy via the use of LMS. The preceding derivation expanded on aspects of an earlier derivation [54], where application of LMS is described in more detail.

Figure 10 shows that the two-dimensional position-change estimation can be completed by calculating the change in heading from

$$\Delta\Psi = \alpha_i - \alpha_j \tag{47}$$

Only one line matched in two different scans is required for this heading-change estimation, but using more lines with LMS improves the accuracy.

Kalman Filter Observables

Eq. (46) can be expressed more generally in vector form as

$$\Delta\rho_k = -\boldsymbol{n}_k \cdot \Delta\boldsymbol{R}_k^{\mathrm{L}} + \Delta\varepsilon_k \tag{48}$$

where $\Delta\rho_k = \rho_k - \rho_{k-1}$ is the change in the range between the laser and an extracted line in the interval between two successive time samples, $\boldsymbol{n}_k = [\cos \alpha_{k-1} \ \sin \alpha_{k-1} \ 0]^{\mathrm{T}}$, $\boldsymbol{R}_k^L$ is the change in the position vector observed in the laser frame, and $\Delta\varepsilon_k$ is the change in measurement errors. At each time step k, the Kalman filter observable z_l^{L} is then the difference between INS estimate of the change in range ($\Delta\rho^{INS}$) and the change in range obtained from the laser ($\Delta\rho$):

$$z_l^{\mathrm{L}} = \Delta\rho_l^{\mathrm{INS}} - \Delta\rho_l \tag{49}$$

where this equation applies for each tracked line l, at each time step k (index not shown), and where

$$\Delta\rho_l^{\mathrm{INS}} = -\boldsymbol{n}_l \cdot \Delta\mathbf{R}^{\mathrm{INS}} + \Delta\boldsymbol{C}_b^{\mathrm{N}}\Delta\boldsymbol{L}_{\mathrm{IMU}}^{\mathrm{L}} \tag{50}$$

with $\Delta\mathbf{R}^{\mathrm{INS}}$ being a change in the position of INS, and $\Delta\boldsymbol{L}_{IMU}^{\mathrm{L}}$ being the lever arm between the IMU and the laser sensor expressed in the body frame.

As from other aiding sensors, z_l^{L} are appended to the measurement vector z and used in the "Kalman Filter" section equations to produce corrections for INS. As in the case of the odometer, the use of INS in manipulation of extracted lines correlates the error in the laser-derived measurements with INS errors. The corresponding Kalman filter covariance matrix terms are continuously updated to reflect this.

CAMERA

Overview

Images captured by a video camera can also be used to extract information that can be used in a Kalman filter to improve positioning. The geometric principles behind the approach are described here. A camera was used as a sensor in combinations examined in the simulation portion of the study but it was not selected for the implementation portion. The aim of this section is to allow conceptual appreciation for the positioning information that can be extracted from video imaging.

An image captured by a camera gives a two-dimensional (2D) representation of a three-dimensional (3D) scene. One image is not sufficient to reconstruct the 3D scene due to lack of depth information and due to obstructions among objects in the scene, but multiple images from different viewpoints can suffice. The multiple images can be captured by several still cameras or by a single moving camera. The latter is the assumption of the approach presented here although it still allows for the use of multiple cameras. Images from a camera moving through a scene can be used to reconstruct the scene and camera motion. The goal here is to estimate the camera motion, which represents the vehicle motion.

Pinhole Frontal Perspective Imaging Model

A thin lens camera model with pinhole aperture and frontal perspective imaging [56] is used here. Figure 11 will be used to illustrate this model. It assumes a lens aperture diameter that is so close to zero as to force all light rays entering the camera to go through a single point called optical center and located in the plane of the lens (focal plane) and marked with O in the figure. In this model an image is formed by an intersection of light rays, which continue through the optical center, and the image plane, which they hit at a distance called focal length, f. Figure 11 shows an image P' formed on the image plane from a point P on an observed object. Point P is located at coordinates (X, Y, Z) with respect to the camera frame x-y-z centered at O. The geometry of the model gives the following relationship between the point object P at (X, Y, Z), its image P' at (X', Y', Z'), and the lens property f:

$$X' = -f\frac{X}{Z}, Y' = -f\frac{Y}{Z} \tag{51}$$

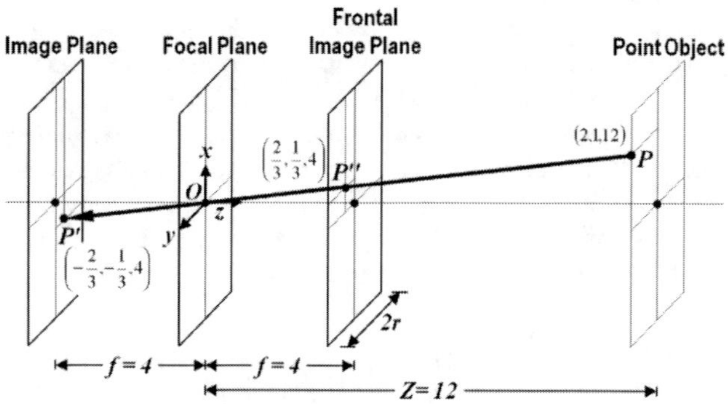

Figure 11. Imaging Model for Pinhole Aperture and Frontal Perspective

The above equation provided the coordinates $-\frac{2}{3}, -\frac{1}{3}, 4$ of the image P′ for the observed point object P at (2,1,12) and the focal length of 4 units.

As the negated image coordinates suggest in Eq. (51), and as the figure shows, the image is inverted in the image plane. The brain compensates for this in the human vision system and it is a simple matter to negate the result in a computer. To avoid the need to negate, the image plane can be imagined to be at a focal length ahead of the focal plane, as represented in the figure with the "Frontal Image Plane". This removes the negative signs from Eq. (51).

Note that Eq. (51) shows that a point Q at (4,2,24) would yield the same image point as point P, $Q' = P'$. This is an example of scale ambiguity: one image can be a result of objects of different sizes and locations. The coordinates of these different objects, expressed in the camera frame, are related by a scale factor, λ. The scale factor between P and Q is $\lambda = 2$.

This last aspect completes the description of the geometry fundamentals of the imaging model used. It should be also noted that photometrically the model assumes a purely refractive lens (no reflection or diffraction), and that all imaged surfaces are Lambertian. The amount of light radiated from a Lambertian surface is not dependent on the vantage point. Given these assumptions, image can be determined purely from ray tracing. An example for a 3D object is shown in Figure 12.

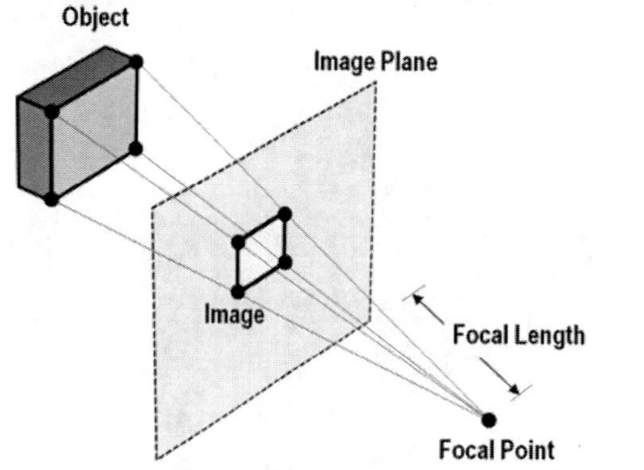

Figure 12. Image of a 3D Object Using Pinhole Frontal Perspective

Unit Sphere Representation

The image surface does not have to be a plane. A unit sphere is used here. It eliminates the problem that arises when using one unified image plane to represent images from multiple cameras [57]. Some cameras' focal planes could be perpendicular to the unified image plane and have objects with light traces that are parallel to the image plane. They would thus offer no projection to the unified plane and equivalently have no representation in such a plane. Even nearly parallel traces are undesirable because small errors in the focal plane transform into large errors projected into the unified plane.

Simplified Illustration: Camera Translation in 2D

Having established the geometrical model of the imaging processes and having chosen a unit-sphere image surface now allows considering extracting motion parameters from the geometry of moving camera images. In order to simply this conceptual representation, a 2D world is used. This means that spheres become circles, planes become lines, and points stay points. Figure 13 shows the motion information that can be extracted from two successive images (at times t_1 and t_2) from a moving camera with its optical center and a unit-circle unified imagining surface both centered at point O. The same subscripts are used to represent other quantities as they change from the first image to the second. An additional simplifying assumption is that the camera moves only in translation (no rotation), as evident from the figure. The figure is complex in order to show the interaction between various representational and analytical aspects, but it can be simple to follow when viewed in stages.

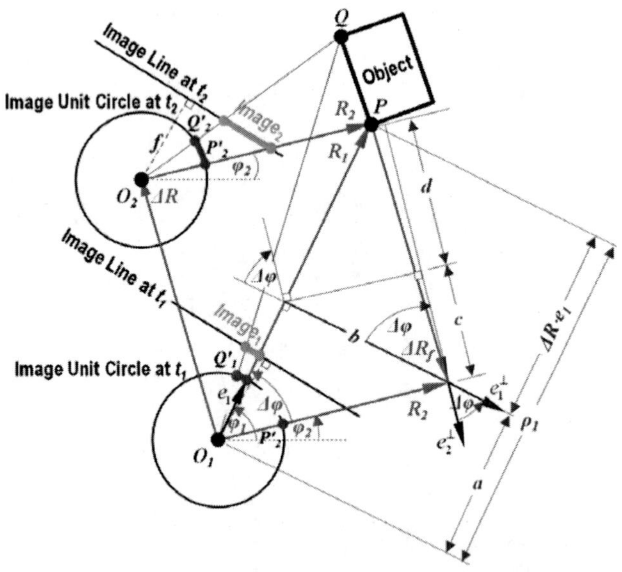

Figure 13. 2D Motion Geometry from Moving Camera Images of Stationary Object

Consider first the single stationary object (the rectangle labeled "Object"). Two points, P and Q, are emphasized on the object. The light traces from those points through the camera's optical center at t_1 and t_2 create respective object images $Image_1$ and $Image_2$ which are shown on the imaging lines by the thick bright blue lines. The unit circle image representations (the green arc segments on the unit circles) are used instead of line representations in the calculations that follow.

Point Features

Being able to recognize image points P' and Q' as coming from geometrical points P and Q on an actual object and track them as they move from one image frame (as P'_1 and Q'_1) to another (as P'_2 and Q'_2) is a significant technical challenge in itself [56]. As the figure shows, their locations in the overall image (on the image line or unit circle in Figure 13) will change as a result of the camera motion. When searching where the point moved, the intensity value of one digital image pixel cannot be used to uniquely represent a point on an object because a typical image has multiple pixels of same intensity. Using values of a window of pixels to present a point helps, but in real images, the values of those pixels will change somewhat as the camera changes the viewpoint to produce the second image (due to real surfaces not always following the Lambertian surface assumption). Despite of these difficulties, research in this area has produced algorithms that can, with some imperfect but acceptable reliability, identify unique points (small windows of pixels) in one image and recognize them after their transformations into the corresponding points in the second image. These uniquely identifiable points are called features.

In Figure 13, the two points P and Q on the object appear as point features P' and Q' which would be identified and tracked using such an algorithm.

Processing of vision in this project was only done in simulation. No actual images were processed to extract and track point features. Instead, new features are simply generated to appear randomly at certain frequency along the trajectory and their migration through images is calculated based on the trajectory. Realism is simulated by adding noise at various steps of the process. Had the vision been used in the implementation, a known feature extraction algorithm, such as SURF [58], could have been used.

Kinematics in Illustration

In Figure 13, only the view of P, from two different locations, is needed to derive the equations. Imagine now only having access to $Image_1$ represented on the unit circle and knowing the camera focal length f. Combined with Eq. (51), this is sufficient to determine the direction, expressed as φ_1, of vector $\boldsymbol{R}_1$. Vector $\boldsymbol{R}_1$ specifies the location of P with respect to camera at t_1. Its magnitude, or range $\rho_1 = \|\boldsymbol{R}_1\|$, is unknown solely from the image geometry due to scale ambiguity. The range can be obtained from other sensors. It can also be found, as assumed here but not detailed, using synthetic stereo vision where camera motion between multiple images is known from INS [57].

Now consider that the camera has moved from O_1 to O_2, as shown in the figure. This gives a new view of P and its new representations in the camera frame as $\boldsymbol{R}_2$ at φ_2. The vector $\boldsymbol{R}_2$ is drawn near the top of the figure in the O_2 frame, but it is also represented shifted down into O_2 frame to show how it can be used with $\boldsymbol{R}_1$ to extract the motion of the camera defined by the vector $\Delta\boldsymbol{R}$, which is the aim of this derivation. The vector $\Delta\boldsymbol{R}$ is equal in magnitude but opposite in direction to the vector $\Delta\boldsymbol{R}_f$, which represents the relative motion of P with respect to O_1 of as seen from O_1. The main geometry of the scene is now set up. All auxiliary geometric quantities in the equations that follow are in the figure also.

Constraint Equation Derivation

The geometry of the figure gives:

$$\Delta\varphi = \varphi_1 - \varphi_2 \tag{52}$$

$$a = \rho_1 - \Delta\boldsymbol{R}\cdot\boldsymbol{e}_1 \tag{53}$$

$$b = -\Delta\boldsymbol{R}\cdot\boldsymbol{e}_1^{\perp} \tag{54}$$

$$a \sin \Delta\varphi = b \cos \Delta\varphi \tag{55}$$

where $\boldsymbol{e}_1$ is the unit vector toward point feature formed from P_1 and $\boldsymbol{e}_1^{\perp}$ is the unit vector perpendicular to $\boldsymbol{e}_1$. Substituting Eq. (52), (53), (54) into (55) and rearranging gives:

$$(\rho_1 - \Delta\boldsymbol{R} \cdot \boldsymbol{e}_1)(\sin \Delta\varphi) = (-\Delta\boldsymbol{R} \cdot \boldsymbol{e}_1^{\perp})(\cos \Delta\varphi)$$
$$(\rho_1)(\sin \Delta\varphi) = (-\Delta\boldsymbol{R} \cdot \boldsymbol{e}_1^{\perp})(\cos \Delta\varphi) + (\Delta\boldsymbol{R} \cdot \boldsymbol{e}_1)(\sin \Delta\varphi) \tag{56}$$

Noting that $\cos \Delta\varphi = \sin(90° - \Delta\varphi)$, $\sin(\Delta\varphi) = \cos(90° - \Delta\varphi)$ and setting $c = (-\Delta\boldsymbol{R} \cdot \boldsymbol{e}_1^{\perp})(\cos \Delta\varphi)$ and $d = (\Delta\boldsymbol{R} \cdot \boldsymbol{e}_1)(\sin \Delta\varphi)$ according to the figure yields:

$$(\rho_1)(\sin \Delta\varphi) = \mathrm{c} + \mathrm{d} \tag{57}$$

The figure also shows that c + d is a projection of $\Delta\boldsymbol{R}$ on $\boldsymbol{e}_2^{\perp}$, thus:

$$(\rho_1)(\sin \Delta\varphi) = -\Delta\boldsymbol{R} \cdot \boldsymbol{e}_2^{\perp} \tag{58}$$

From Eq. (58) it is obvious that the desired camera displacement $\Delta\boldsymbol{R}$ is only a function of the known image properties $\boldsymbol{e}_2^{\perp}$ and $\Delta\varphi$, and range ρ_1, which is only required for the first image. This constraint equation can be used with subsequent images and an initial range to continually estimate the camera position.

Expanding the formulation to 3D with both translational and rotational camera motions produces a set of constraint equations Eq. (59) [57].

$$(\boldsymbol{e}_2)^{\mathrm{T}} \Delta\boldsymbol{C}_N^b \boldsymbol{B} \Delta\boldsymbol{R} = (\boldsymbol{e}_1)^{\mathrm{T}} \boldsymbol{B}^{\mathrm{T}} \Delta\boldsymbol{C}_b^N \boldsymbol{e}_2 \rho_1$$
$$(\boldsymbol{e}_2)^{\mathrm{T}} \Delta\boldsymbol{C}_N^b \boldsymbol{D} \Delta\boldsymbol{R} = (\boldsymbol{e}_2)^{\mathrm{T}} \Delta\boldsymbol{C}_N^b \boldsymbol{e}_2^{\perp} \rho_1 \tag{59}$$

where $B = \begin{bmatrix} 0 & -1 & 0 \\ 1 & 0 & 0 \\ 0 & 0 & 0 \end{bmatrix}$ and $D = \begin{bmatrix} 0 & 0 & -\cos\varphi_1 \\ 0 & 0 & -\sin\varphi_1 \\ \cos\varphi_1 & \sin\varphi_1 & 0 \end{bmatrix}$.

Kalman Filter Observables

A Kalman filter observable for each point feature p in each image 1 and n is then formed according to Eq. (60) [57].

$$\eta_n^p = \begin{bmatrix} (\boldsymbol{e}_n^p)^{\mathrm{T}} \Delta\boldsymbol{C}_N^b \boldsymbol{B} \Delta\boldsymbol{R}_{\mathrm{INS}} - (\boldsymbol{e}_1^p)^{\mathrm{T}} \boldsymbol{B}^{\mathrm{T}} \Delta\boldsymbol{C}_b^N \boldsymbol{e}_n^p \rho_1^p \\ (\boldsymbol{e}_n^p)^{\mathrm{T}} \Delta\boldsymbol{C}_N^b \boldsymbol{D} \Delta\boldsymbol{R}_{\mathrm{INS}} - (\boldsymbol{e}_n^p)^{\mathrm{T}} \Delta\boldsymbol{C}_b^N \boldsymbol{e}_{\perp}{}_n^p \rho_1^p \end{bmatrix} \tag{60}$$

In Eq. (60) $\Delta\boldsymbol{R}_{\mathrm{INS}}$ and $\Delta\boldsymbol{C}_b^N$ are INS-derived body position and orientation changes between images 1 and n. Equation (60) can also be linearized [57].

SENSOR SELECTION VIA SIMULATION

Simulation was created to evaluate several sensor combinations before selecting two for vehicle implementation.

The five sensor combinations evaluated were GPS/INS (GI for short), GPS/INS/Odometer (GIO), GPS/Clock/INS/Odometer (GCIO), GPS/Clock/INS/Odometer/Camera (GCIOC) and GPS/Clock/INS/Odometer/Lidar (GCIOL).

Actual satellite geometry was used for simulated test durations. The sensors' error characteristics were modeled using noise added to true geometry based on sensor noise statistics. The simulation allowed accounting for the effect of the test environment by modeling the satellite obstructions created by the test route surroundings (Figure 14). The building height and street width information (Figure 15) was extracted from fire insurance maps [59].

The version of algorithm used was very similar to that applied later to actual data. The performance was evaluated for several visibility scenarios: 3D environment model, 3 satellites (space vehicles, SVs), 2SVs, 1 SVs, and 0 SVs. Each simulated test run included a 1 minute initialization portion that simulates building obstructions and was followed by 9 minutes of continued simulated driving along the route for the scenarios with 2 SVs or more, and shorted durations for other scenarios. There were 3 test runs per sensor combination. The test runs were separated in time to simulated the effect of changing satellite geometry. Figure 16 is an example test run. It is for GCIOL in complete outage conditions.

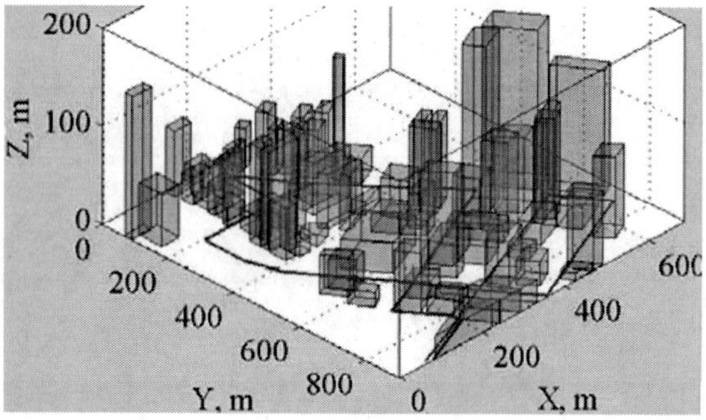

Figure 14. 3D Model of Test Environment

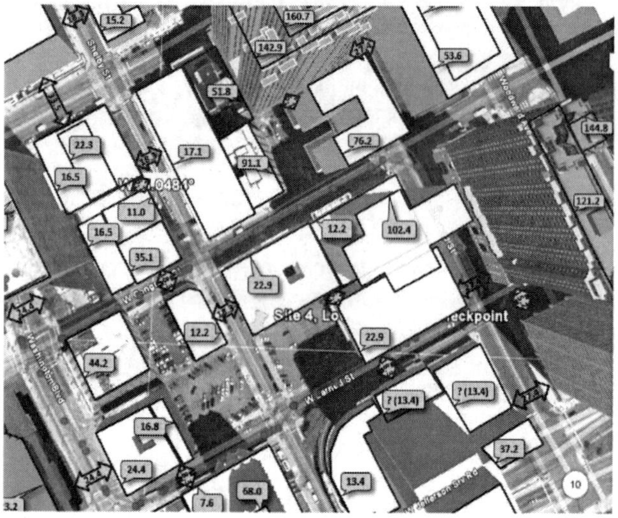

Figure 15. Building Heights and Street Widths Model, Downtown Detroit

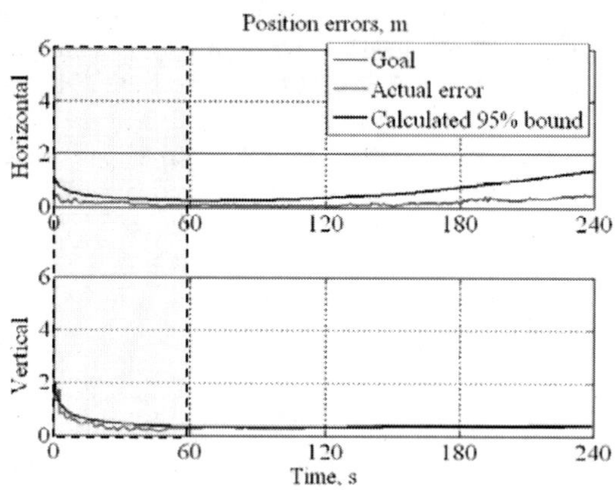

Figure 16. Example Simulation Output

The summary of the simulation results appears in Table 2. Options that that were found unsuccessful in the easier scenarios (left two columns in Table 2) were not simulated for all scenarios. Those options are not shown in Table 2. The table shows that the GCIOL option performs best. It meets the requirements in all simulated environments except one (the exception is the 1 SV, 5 min case). Thus GCIOL option was selected as one of the two options for implementation. GCIO was selected as the second option because: (1) it meets the requirement for the simulated environment as do the other options in Table 2, (2) its sensors are already part of GCIOL, and (3) when it fails to meet the requirement over the entire duration of the three outage tests (right three columns of Table 2) it can still meet it for shorter durations (< 70 s).

Table 2. Simulation Results

First row – maximum position error over example simulation runs
Second row – 95% error bound calculated based on covariance of the Kalman filter

Option \ Scenario	Detroit urban canyon, 10 min	Difficult GPS environment, 3SV case, 10 min	Difficult GPS environment, 2SV case, 10 min	GPS outage, 1SV case, 5 min	Complete GPS outage, 3 min
GIO	0.5 m (max) 1 m (max)	1.8 m (max) 1.6 m (max)	7.5 m (max) 11 m (max), <2 m if 100 sec	10 m (max) 18 m (max) <2 m if t<70 sec	2.5 m (max) 8 m (max) <2 m if t<70 sec
GCIO*	< 0.5 m < 1 m	< 1.8 m < 1.6 m	7.5 m (max) 7.5 m (max), <2 m if t<100 sec	3.2 m (max) 6 m (max), <2 m if t<80 sec	2.5 m (max) 8 m (max) <2 m if t<70 sec
GCIOC	< 0.5 m < 1 m	< 1.8 m < 1.6 m	2.1 m (max) 2.2 m (max) <2 m if t<100 sec	2.5 m (max) 3 m (max) <2 m if t<80 sec	2.1 m (max) 2.1 m (max) <2 m if t<90 sec
GCIOL	< 0.5 m < 1 m	< 1.8 m < 1.6 m	1 m (max) 2 m (max)	2 m (max) 3 m (max) <2 m if t<100 sec	1.7 m (max) 1.7 m (max)

**Does not include simulation over multiple runs*

VEHICLE IMPLEMENTATION

SENSOR ACQUISITION SYSTEM

Sensor interfacing and logging was done using a National Instruments (NI) chassis (NI PXI 1042-Q) filled with a real-time computer (NI PXI-8108) and NI serial and digital I/O cards. The computer had an Intel 2.53 GHz dual core T9400 processor with 1 GB RAM. The system was programmed using LabVIEW RT graphical programming language.

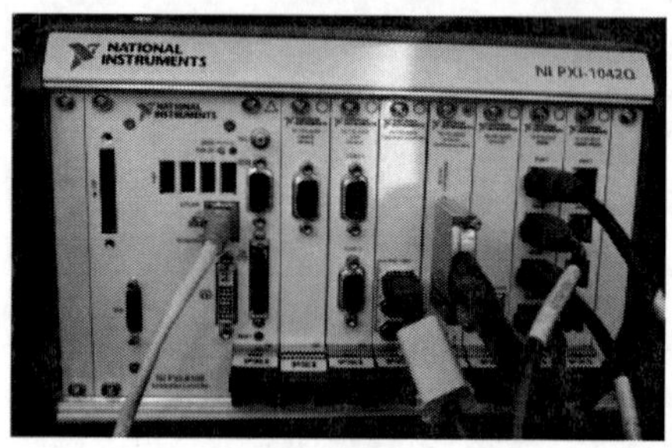

Figure 17. Sensor Acquisition System

Data was collected during downtown test driving and afterwards passed through the positioning algorithm running in Matlab™. Data was logged and applied to the algorithm in a manner that simulates real-time operation (i.e., at every time step the algorithm could only access the data available at that time during testing). This emphasized real-time implementation challenges, such as the fact that the algorithm expects the GPS data to be available immediately at each 10 ms time step while in reality in arrived 50 to 300 ms later.

All data was tagged with GPS times using information from the GPS receiver. The pulse-per-second (PPS) signal from the GPS receiver was used to trigger a software interrupt task that recorded the CPU time at the PPS rising edge ($PPS_{t_{CPU}}$). The PPS edge was also associated with the GPS time ($PPS_{t_{GPS}}$) relayed in the GPS timing message that followed the PPS edge. Then any measurement event in the following second was tagged with GPS time as follows:

$$event_{t_{GPS}} = PPS_{t_{GPS}} + (event_{t_{CPU}} - PPS_{t_{CPU}}) \tag{61}$$

The accuracy of this approach for assigning a time to another incoming edge (for example the odometer edge) was measured to be ±25 μs with respect to the receiver's GPS time. The uncertainty mainly comes from the variability in task triggering. It is expected that using a simple custom operating system on a fast microcontroller could reduce this to the order of ns. There was a hardware filter added to the PPS signal to exclude noise, but its delay is constant and thus was removed. In addition, there were several levels of software logic to prevent incorrect timing due to serial communication or PPS signal problems.

The IMU and laser measurements came over serial communication without timestamps and the only discernable event available for time measurement was the actual arrival of serial messages. Using the arrival of serial messages for timestamping introduced another 300 μs of uncertainty. This was still deemed to be acceptable for the algorithm.

INS

The IMU for the proposed system (small device shown in Figure 18) and the reference device IMU (large device shown in Figure 18) were both mounted securely on the floor of the test vehicle at a central location. They were in close horizontal proximity to their corresponding GPS antennas, which were on the roof the test vehicle vertically 1.3 m away.

The IMU for the proposed system provided its measurements at a rate of 450 Hz via serial communication at 115.2 kBd. For efficiency, only every third set of measurements was used for an effective rate of 150 Hz. No pre-filtering was required (but note that the positioning algorithm uses a Kalman filter). The measurements were linear accelerations in m^2/s along three axis and angular speeds around three axis in °/s. The manufacturer specifies the linear acceleration in-run bias stability to be ≤ 3 mg and the angular speed in-run bias stability to be ≤ 100°/h.

Figure 18. IMU Installation

GPS

The GPS receiver was mounted in the trunk and the antenna on the vehicle roof (Figure 19). The GPS output was provided to the sensor acquisition system at the rate of 1 Hz over 921.6 kBd serial communication.

Figure 19. Test Vehicle Roof Installations

The GPS data arrived between 50 and 350 ms after the time to which it corresponds. Since the data was logged and used in a way that simulates real-time operation, this created a problem for the positioning algorithm, which expects GPS data to be available within the 10 ms period from the IMU data. A buffer was created in the positioning algorithm to eliminate this problem. It operates by suspending positioning processing and storing other sensor values (IMU, odometer, laser) until the required GPS data arrives and then catching up by calculating several delayed execution steps within one actual execution step.

OSCILLATOR

The signal from the oscillator is interfaced directly with the receiver, as provided by the receiver. The receiver is configured to use this signal instead of its built-in oscillator. Thus, the oscillator signal is not processed by the sensor acquisition component of the proposed positioning system. The use of the oscillator only affects the initial clock error statistics assumptions applied to the Kalman filter.

ODOMETER

Vehicle odometry is derived from the vehicle speed signal generated by the powertrain control module, which calculates vehicle speed based on sensing the speed of the engine output shaft and generating a representative signal. Measurement showed that a rising edge is generated on average about 5.5 times per wheel revolution. Pulses were observed to truly coincide with the incremental angular position of the wheel even at near-zero speeds, which makes them useable for this approach. A signal merely proportional to the vehicle speed without pulse synchronization with wheel angular position would not be effective.

This particular signal was used due to ease of access and its discrete nature (either high or low), which allowed precise timing. In contrast, obtaining measurements over a vehicle communication network, such as CAN, could introduce timing uncertainties that would be more detrimental than low resolution or occasional wheel slipping. Using wheel sensors at each wheel would give a less ambiguous measurement than those based on the output shaft, provided they can be accessed safely and timed precisely. This measurement was determined to generally represent speed of the vehicle at the middle of the front axle. The all-wheel drive configuration of the test vehicle greatly reduced the chances of wheel slipping.

LASER

The laser was mounted on the roof of the vehicle facing forward (Figure 19, Figure 20). This location allowed horizontal scanning that reached surrounding city architecture without being significantly affected by traffic obstructions.

Figure 20. Test Vehicle

The interfacing with the laser was also via serial communication. The laser sends 374 bytes of measurement data at 75 Hz over serial communication at 38.4 kBd, but data was only kept at 1 Hz. The arrival of each serial message was used for time-stamping. Since the laser takes 13.3 ms to perform one scan which consists of 181 angular measurements, assigning one time-stamp to all the measurements would introduce delay up to 13.3 ms that varies with the measurement angle, which is not acceptable for the algorithm. Timing corrections for each angular measurement were implemented to remove this effect.

EXPERIMENTAL EVALUATION

TESTING METHOD

The two selected solutions, GCIO and GCIOL, were tested in the sky-obstructed environment of downtown Detroit (Figure 21). The test route consisted of a 2 km long, mostly open-sky approach (not in Figure 21), followed by the 2.3 km difficult-conditions loop shown in Figure 21. The error analysis was performed only for the difficult conditions loop. Although the test environment features streets lined with buildings of various heights, including some partially open areas, the masking angle formed between the middle of the average street and the top of the average building is very high at 75°. The test procedure [60] generates the truth reference as combination of geodetic survey data and post-processed data from an in-vehicle reference-grade GPS/INS system.

Figure 21. Test Environment with Test Route Shown

A distinction is made between the summed duration of all valid test runs in one day and the time spanned from the start of the first valid test run to the end of the last valid test run. The first is referred to as the total test driving duration and it serves to quantify the actual amount of data collected. The second, referred to here as span duration, gives the indication of the amount of satellite geometry variability captured during the test day. The procedure [60] suggests spanning 24 h because the satellite ground tracks repeat approximately once every 24 h. The total testing performed spanned 38.5 h, however, some of it is not usable for performance analysis due to work on improving the data acquisition system and the test procedure. Table 3 presents test days with useable data. All test days logged the data for GPS-only option. There were test days with both GCIO and GCIOL on-board and test days with only GCIO. The 22.5 h span of testing that involves GCIO almost reaches the goal of 24 h, but GCIO+GCIOL testing only spans 8 h. Due to limited GCIO+GCIOL test time, in order to ensure the same test conditions when comparing the performance of GCIO and GCIOL options, only GCIO+GCIOL testing will be used for that purpose. All of GCIO testing will be used for the assessment of GCIO to take advantage of the desired additional variability coverage.

Table 3. Testing Overview

2D horizontal 95 % error cumulative for entire test day in 2.3 km of difficult conditions
avail % availability of solutions with error less than 2 m
ADF difficulty factor, average over all runs
- data not available
Dur. duration

calendar date	day label	logged ("L"), analyzed (2D/avail) A	GCIO	GCIOL	ADF (%)	num.	valid test runs durations avg. (min)	all (h)	span (h)
2010-02-11	D02	78.8/34	2.3/90	-	51	7	18	2:07	6:09
2010-03-19	D04	49.5/41	4.8/72	3.8/80	67	5	16	1:18	5:37
2010-05-17	D05	24/35	5.3/91	2.6/89	45	3	16	0:48	2:20
2010-09-02	D07	33.5/38	4.3/89	-	59	7	21	2:22	8:27
Device Totals:		**58.6/37**	**4.1/86**	**3.4/85**	**56**	**22**	**18**	**6:35**	**22:33**
ADF		56	56	56	**Testing Totals**				
Valid Runs		6	22	8					
Test Driving (h)		6:35	6:35	2:06					
Test Span (h)		22:33	22:33	7:57					

Total for Tests with A, GCIO, and GCIOL	A	GCIO	GCIOL
2D	**44.2**	**4.7**	**3.3**
avail	**38**	**82**	**85**
ADF	56	56	56
Valid Runs	8	8	8
Test Driving (h)	2:06	2:06	2:06
Test Span (h)	7:57	7:57	7:57

GPS

The results for the GPS-only device, the same device whose GPS data is used in GCIO and GCIOL, are shown in Table 3, column A. The results are presented as pairs of error and availability. For example, the total for all GPS testing is 58.6/37. This represents 58.6 m 95% horizontal error and 37% availability of errors less than 2 m. These values are cumulative for testing on all relevant days which are in this case D02, D04, D05, and D07.

This device is expected to perform poorly in this environment. It performance assessment is included here only as a example of current representative unaided performance. This illustrates the need for solutions such as GCIO and GCIOL.

GPS/CLOCK/INS/ODO (GCIO)

The assessment of GCIO performance is 4.1/86 (device totals row and the GCIO column of Table 3). This reflects a test span of 22.5 h and is thus expected to be a fair representation of this device in this environment. Notice the dramatic improvement over the baseline GPS-only case. The GCIO solution overall approaches the sought after 2 m accuracy and in a number of individual test runs exceeds it (Table 4) suggesting that there are difficult environments in which it would meet the requirements.

Table 4. Test Day D02 Error Assessment Summary for GCIO

Test Run	Difficulty Factor (%)	Across Accuracy (vs. sur.) (95 %)	Along Accuracy (vs. ref.) (95 %)	2D Horizontal Accuracy (vs. surv./ref.) (95 %)	Vertical Accuracy (vs. ref.) (95%)	Availability (vs. surv./ref.) (H. Error <= 2 m) [%]
1	60	1.7	1.3	2.2	4.3	96
2	61	1.7		1.7		90
3	69	1.3	2.0	2.2	2.8	89
4	54	1.8	1.8	2.6	5.2	80
5	35	2.2	1.6	2.6	2.7	94
6	31	1.2	2.0	2.7	7.4	83
7	46	1.3	1.1	1.5	3.6	100
Average	51	**1.6**	**1.6**	**2.2**	**4.3**	**90**

GPS/CLOCK/INS/ODO/LASER (GCIOL)

GCIOL scores 3.4/85 although its test data is only from 2.5 h of actual test driving and spans only 8 h of variability. This allows the possibility that encountering more diverse conditions could have affected the numbers. GCIOL performance is still closer to the goal of 2 m. As the second number in the pair shows, the goal is met 85% of the time.

GCIO VERSUS GCIOL

Comparing GCIO and GCIOL using testing that involves both shows that GCIOL reduces error by 1.4 m. However the improved accuracy shown by GCIO in extended testing suggests that the difference is maybe closer to 1 m. More testing is required for more confidence but there are many individual GCIOL test runs that perform better than GCIO while the rest are similar (Figure 22).

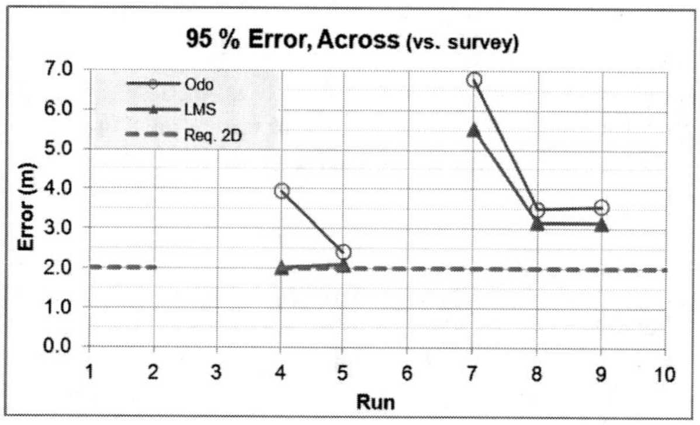

Figure 22.

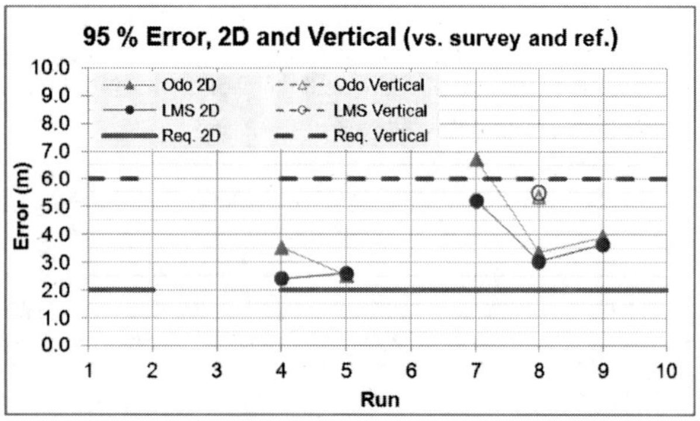

Figure 22 (cont.). Test Day D04 Errors for GCIO ("Odo" in legend) and GCIOL ("LMS")

QUALIFICATIONS AND SUGGESTED IMPROVEMENTS

Several qualifications are in order that lead to suggested improvements.

1. In order to account for the influence of the interaction of satellite geometry changes and the environment, the test durations suggested by the test procedure [60] should be used as minimums. This was not achieved here. As a result, there is some uncertainty in the overall 95% error assessment values. We speculate that overall 95% error assessment values may be off by as much ±0.5 m compared to a testing regiment that more closely follows the procedure.

2. The system used for sensor data acquisition and time-stamping was not fully tested. It was undergoing minor improvements throughout the testing. Precise timing is critical for the performance of the positioning algorithm and so it is expected that this may have caused some performance degradation.

3. This performance is a result of taking an algorithm from desk to road with minimal adaptations, thus further improvements are expected to be achievable.

Based on the above qualifications, it is suggested that the proposed positioning system, after its further development, is retested with more disciplined adherence to the testing procedure [60].

SUMMARY

A multi-sensor positioning system is proposed that increases the availability of lane-level accuracy in dense urban environments, which can often make standalone GPS ineffective. An algorithm based on the Kalman filter that uses data from GPS, INS, odometer, laser, and camera sensors is described with attention given to the processing peculiarities of each sensor. A simulation is used to limit the combinations of sensors examined to two: GPS/Clock/INS/Odometer and GPS/Clock/INS/Odometer/Laser. The selected combinations are instrumented onto a test vehicle and data collected during test driving in an actual difficult-GPS environment. The data is passed through the positioning algorithm to assess the positioning performance of the two options. In difficult conditions only, the GPS/Clock/INS/Odometer option achieves 95% error of 4.1 m and provides 2 m accuracy 86% of the time, while the limited testing of GPS/Clock/INS/Odometer/Laser solution shows a 95% error of 3.4 m and 2 m availability of 85%.

Future work should involve improvements in the reliability of sensor data acquisition timing and further adaptation of the algorithm to real-time peculiarities. This is to be followed by retesting to assess the expected performance improvement. The testing should more closely follow the intended test procedure, specifically with respect to longer test duration.

REFERENCES

1. U.S. Department of Transportation, "An Analysis of the Significant Decline in Motor Vehicle Traffic Fatalities in 2008," June 2010.

2. U.S. Department of Transportation, "2008 Traffic Safety Annual Assessment - Highlights," June 2009.

3. Maile, M. and Delgrossi, L., "Cooperative Intersection Collision Avoidance System for Violations (CICAS-V) for Avoidance of Violation-Based Intersection Crashes," National Highway Traffic Safety Administration, Washington, DC.

4. Maile, M., Ahmed-Zaid, F., Basnyake, C., Caminiti, L., Kass, S., Losh, M., Lundberg, J., Masselink, D., McGlohon, E., Mudalige, P., Pall, C., Peredo, M., Popovic, Z, Stinnett, J., and VanSickle, S. (In Print), "Cooperative Intersection Collision Avoidance System Limited to Stop Sign and Traffic Signal Violations (CICAS-V) Task 11 Final Report: Objective Tests," National Highway Traffic Safety Administration, Washington, DC.

5. Ahmed-Zaid, F., Bai, F., Bai, S., Basnayake, C., Bellur, B., Brovold, S., Brown, G., Caminiti, L., Cunningham, D., Elzein, H., Ivan, J., Jiang, D., Kenny, J., Krishnan, H., Lovell, J., Maile, M., Masselink, D., McGlohon, E., Mudalige, P., Rai, V., Stinnett, J., Tellis, L., Tirey, K., VanSickle, S. "Vehicle Safety Communications - Applications (VSC-A) First Annual Report - December 7, 2006 through December 31, 2007," National Highway Traffic Safety Administration, Washington, DC, Sept. 2008.

6. "IEEE 802.11p: Towards an International Standard for Wireless Access in Vehicular Environments," presented at Vehicular Technology Conference, USA, May 2008.

7. ASTM, "Standard Specification for Telecommunications and Information Exchange Between Roadside and Vehicle Systems - 5 GHz Band Dedicated Short Range Communications (DSRC) Medium Access Control (MAC) and Physical Layer (PHY) Specifications," ASTM E2213 - 03(2010), doi:10.1520/E2213-03R10, 2010.

8. IEEE, "Wireless LAN Medium Access Control (MAC) and Physical Layer (PHY) Specifications," IEEE Press, New York, NY, February, 2007.

9. SAE Surface Vehicle Standard Draft, "Dedicated Short Range Communications (DSRC) Message Set Dictionary," SAE Standard Draft J2735, May 2008.

10. "Understanding GPS: Principles and Applications, 2nd ed.," edited by Hegarty, Christopher and Elliott, D., Norwood, MA: Artech House, 2006.

11. Misra, P., and Enge, P., "Global Positioning System: Signals Measurements, and Performance, 2nd ed.," Ganga-Jamuna Press, Lincoln, MA, 2006.

12. Massat, P. D., "GNSS Sustainment: The Challenge of Availability," plenary session at ION GNSS 2010 conference, Portland, OR, 2010.

13. ARINC Engineering Services, "NAVSTAR GPS Space Segment/Navigation User Interfaces, Interface Specification IS-GPS-200D," U.S. Air Force, Mar. 2006.

14. U.S. Department of Defense, "Global Positioning System Standard Positioning Service Performance Standard," 2001.

15. u-blox, "u-blox 5 GPS and GALILEO Receiver Module with Precision Timing," LEA-5T Brochure, 2008.

16. NovAtel OEMV-3 Brochure, "Multi-Frequency GNSS Receiver Provides Expandable Functionality Without Compromising Performance," 2009.

17. Shah, R., Nowakowski, C., Green, P. "U.S. Highway Attributes Relevant to Lane Tracking," Technical Report UMTRI-98-34, Aug. 1998.

18. Southworth, M., and Ben-Joseph, E., "Street Standards and the Shaping of Suburbia," *APA Journal*, 66-81, Winter 1995.

19. Basnayake, C., Williams, T., Alves, P., and Lachapelle, G., "Can GNSS Drive V2X?" GPS World, Oct. 2010.

20. Federal Highway Administration, "Annual Vehicle Miles of Travel, 1980-2007, By Functional System, National Summary (Table VM-202)," http://www.fhwa.dot.gov/policyinfomation/statistics/vm02_summary.cfm, Jan. 2009.

21. Xiaohui, B., Haiyang, L., Rui, Z., Jie, C., "A Novel Algorithm Based on FFT for Ultra High-Sensitivity GPS Tracking," *ION GNSS 2009 Conference Proceedings*, Savannah, GA, Sept. 2009.

22. Iwase, T., Suzuki, N., Watanabe, Y., "Estimation of Multipath Range Error for Detection of Erroneous Satellites," *ION GNSS 2010 Conference Proceedings*, Portland, OR, Sept. 2010.

23. Mattos, P. G., "Adding Glonass to the GPS/Galileo consumer receiver, with hooks for Compass," *ION GNSS 2010 Conference Proceedings*, Portland, OR, Sept. 2010.

24. O'Driscoll, C., Tamazin, M. E., Lachapelle, G., "Investigation of the Benefits of Combined GPS/GLONASS for High Sensitivity Receivers," *ION GNSS 2010 Conference Proceedings*, Portland, OR, Sept. 2010.

25. Li, T., Petovello, M. G., Lachapelle, G., Basnayake, C., "Performance Evaluation of Ultra-tight Integration of GPS/Vehicle Sensors for Land Vehicle Navigation," *ION GNSS 2009 Conference Proceedings*, Savannah, GA, Sept. 2009.

26. Karamat, T., Georgy, J., Iqbal, U., Noureldin, A., "A Tightly-Coupled Reduced Multi-Sensor System for Urban Navigation," *ION GNSS 2009 Conference Proceedings*, Savannah, GA, Sept. 2009.

27. Ye, P., Du, G., Zhan, X., Zhai, C., "Performance Evaluation of Compact MEMS IMU/GPS Tight Coupling with IMU Aided Tracking Loop," *ION GNSS 2010 Conference Proceedings*, Portland, OR, Sept. 2010.

28. Somieski, A., Hollenstein, Ch., Favey, E., Schmid, C., "Low-Cost Sensor Fusion Dead Reckoning using a Single-Frequency GNSS Receiver Combined with Gyroscope and Wheel Tick Measurements," *ION GNSS 2010 Conference Proceedings*, Portland, OR, Sept. 2010.

29. Wang, J. J., Kodagoda, S., Dissanayake, G., "Vision Aided GPS/INS System for Robust Land Vehicle Navigation," *ION GNSS 2009 Conference Proceedings*, Savannah, GA, Sept. 2009.

30. Sahmoudi, M., Calmettes, V., Barraut, V., Priot, B., "A New Approach for Ultra-Tight Integration of GNSS and Vision-Aided MEMS IMU," *ION GNSS 2010 Conference Proceedings*, Portland, OR, Sept. 2010.

31. Alam, N., Balaei, A.T., Dempster, A.G., "Positioning Enhancement with Double Differencing and DSRC," *ION GNSS 2010 Conference Proceedings*, Portland, OR, Sept. 2010.

32. Allen, J., and Bevly, D., "Performance Evaluation of Range Information Provided by Dedicated Short-Range Communication (DSRC) Radios," *ION GNSS 2010 Conference Proceedings*, Portland, OR, Sept. 2010.

33. Petovello, M. G., O'Keefe, K., Chan, B., "Demonstration of Inter-Vehicle UWB Ranging to Augment DGPS for Improved Relative Positioning," *ION GNSS 2010 Conference Proceedings*, Portland, OR, Sept. 2010.

34. Draganov, S., Harlacher, M., Haas, L., "Multipath Mitigation via Synthetic Aperture Beamforming," *ION GNSS 2009 Conference Proceedings*, Savannah, GA, Sept. 2009.

35. Attia, M., Moussa, A., El-Sheimy, N., "Bridging Integrated GPS/INS Systems with Geospatial Models for Car Navigation Applications," *ION GNSS 2010 Conference Proceedings*, Portland, OR, Sept. 2010.

36. Sugimoto, S., Kubo, Y., Tanikawara, M., "A Review and Applications of the Nonlinear Filters to GNSS/INS Integrated

Algorithms," *ION GNSS 2009 Conference Proceedings*, Savannah, GA, Sept. 2009.

37. Saeedi, S., El-Sheimy, N., Syed, Z., "The Merits of UKF and PF for Integrated INS/GPS Navigation Systems," *ION GNSS 2009 Conference Proceedings*, Savannah, GA, Sept. 2009.

38. Aggarwal, P., and El-Sheimy, N., "Hybrid Extended Particle Filter (HEPF) for INS/GPS Integrated System," *ION GNSS 2009 Conference Proceedings*, Savannah, GA, Sept. 2009.

39. Lee, A., "Use of Neural Network for the Improvement of Particle Filter Performance in INS/GPS Integrated Navigation System During GPS Signal Outages," *ION GNSS 2010 Conference Proceedings*, Portland, OR, Sept. 2010.

40. Iqbal, U., Georgy, J., Korenberg, M.J., Noureldin, A., "Modeling Residual Errors of GPS Pseudoranges by Augmenting Kalman Filter with PCI for Tightly-Coupled RISS/GPS Integration," *ION GNSS 2010 Conference Proceedings*, Portland, OR, Sept. 2010.

41. Lee, J. K., Toth, C., Grejner-Brzezinska, D.A., "Network-Based Collaborative Navigation for Ground-Based Users in GPS-Challenged Environments," *ION GNSS 2010 Conference Proceedings*, Portland, OR, Sept. 2010.

42. Soloviev, A., Popovic, Z., Mochizuki, Y., "High Precision Positioning in Difficult GPS Environments for Cooperative Vehicle Safety Applications," *ION GNSS 2010 Conference Proceedings*, Portland, OR, Sept. 2010.

43. Soloviev, A., and Miller, M., "Navigation in Difficult Environments: Multi-Sensor Fusion Techniques," NATO RTO Lecture series, Spring 2010.

44. Kalman, R. E., "A New Approach to Linear Filtering and Prediction Problems," *Trans. ASME - J. Basic Eng.* 35-45, March 1960.

45. Kalman, R. E., and Bucy, R., "New Results in Linear Filtering and Prediction," *Trans. ASME - J. Basic Eng.* 83, 95, 1961.

46. Welch, G., and Bishop, G., "An Introduction to the Kalman Filter," University of North Carolina at Chapel Hill, Department of Computer Science, SIGGRAPH 2001 Course 8 course notes.

47. Maybeck, P. S. "Stochastic Models, Estimation, and Control," Vol. 1, New York: Academic Press, 1979.

48. Brown, R. G., and Hwang, P. Y. C., "Introduction to Random Signals and Applied Kalman Filtering," New York: John Wiley & Sons, 1997.

49. The Analytic Sciences Corporation, "Applied Optimal Estimation," edited by Gelb, Arthur, Cambridge, MA: MIT Press, 1974.

50. Titterton, D. H., and Weston, J. L., "Strapdown Inertial Navigation Technology, Second Edition," Reston, VA: The American Institute of Aeronautics, 2004.

51. Van Graas, F., Soloviev, A., "Precise Velocity Estimation Using a Stand-Alone GPS Receiver," *NAVIGATION, Journal of the Institute of Navigation* **51**(4), 2004.

52. Golub, G.H., and Van Loan, C. F., "Matrix Computations," 2nd ed., Baltimore, MD: The John Hopkins University Press, 1989.

53. Wenzel Associates, Inc., "Small Fry OCXO," Datasheet. http://www.wenzel.com/pdffiles1/Oscillators/SF.pdf.

54. Soloviev, A., Bates, D., Van Graas, F., "Tight Coupling of Laser Scanner and Inertial Measurements for a Fully Autonomous Relative Navigation Solution," *NAVIGATION, Journal of the Institute of Navigation* **53**(3), 2007.

55. Spiegel, M. R., Liu, J., "Mathematical Handbook of Formulas and Tables, Second Edition," New York: McGraw-Hill, 1999.

56. Ma, Y., Soatto, S., Kosecka, J., Sastry, S., "An Invitation to 3-D Vision," New York: Springer, 2004.

57. Soloviev, A., Touma, J., Klausutis, T., Miller, M., Rutkowski, A., Fontaine, K., "Integrated Multi-Aperture Sensor and Navigation Fusion," *ION GNSS 2009 Conference Proceedings*, Savannah, GA, Sept. 2009.

58. Bay, H., Tuytelaars, T., Gool, L. V., "SURF: Speeded up robust features," *Computer Vision - ECCV 2006* Lecture Notes in Computer Science **3951**:404-417, 2006, doi: 10.1007/11744023_32.

59. Sanborn Mapping and Geographic Information Services, "Sanborn Fire Insurance Maps for Detroit, MI, 1991 Edition," Pelham, NY, 1991.

60. Popovic, Z. and Mochizuki, Y., "A Method for Testing GPS in Obstructed Environments Where GPS/INS Reference Systems Can Be Ineffective," SAE Technical Paper 2011-01-1036, 2011, doi:10.4271/2011-01-1036.

CONTACT INFORMATION

Questions and suggestions are welcome. Please direct all correspondence to:

Zeljko Popovic
Honda R & D Americas, Inc.
Automobile Technology Research
1000 Town Center Suite 2400
Southfield, MI 48075
248-304-4703
zpopovic@oh.hra.com

ACKNOWLEDGMENTS

The authors would like to thank Radovan Miucic for assistance in this study.

The Engineering Meetings Board has approved this paper for publication. It has successfully completed SAE's peer review process under the supervision of the session organizer. This process requires a minimum of three (3) reviews by industry experts.

ISSN 0148-7191

SAE Customer Service:
Tel: 877-606-7323 (inside USA and Canada)
Tel: 724-776-4970 (outside USA)
Fax: 724-776-0790
Email: CustomerService@sae.org
SAE Web Address: http://www.sae.org
Printed in USA

SAE International

SAE International®

Vehicle Safety Communications - Applications: Multiple On-Board Equipment Testing

2011-01-0586
Published
04/12/2011

Farid Ahmed-Zaid
Ford Motor Company

Hariharan Krishnan
General Motors Company

Michael Maile
Mercedes Benz REDNA

Lorenzo Caminiti
Toyota Motor Engineering & Mfg NA Inc.

Sue Bai
Honda R&D Americas Inc.

Joseph Stinnett
Ford Motor Company

Steve VanSickle
Danlaw, Inc.

Drew Cunningham
Toyota Technical Center USA Inc.

doi:10.4271/2011-01-0586

ABSTRACT

The United States Department of Transportation (USDOT) and the Crash Avoidance Metrics Partnership-Vehicle Safety Communications 2 (CAMP-VSC2) Consortium (Ford, General Motors, Honda, Mercedes-Benz, and Toyota) initiated, in December 2006, a three-year collaborative effort in the area of wireless-based safety applications under the Vehicle Safety Communications-Applications (VSC-A) Project. The VSC-A Project developed and tested Vehicle-to-Vehicle (V2V) communications-based safety systems to determine if Dedicated Short Range Communications (DSRC) at 5.9 GHz, in combination with vehicle positioning, would improve upon autonomous vehicle-based safety systems and/or enable new communications-based safety applications.

A crucial element required for potential deployment of V2V safety systems is the understanding of how DSRC will perform as larger numbers of DSRC radios are added to the system and ensuring that the communication channel can support a large number of vehicles in potentially congested traffic conditions. This is referred to as system scalability. In the VSC-A Project, a preliminary, multiple On-Board Equipment (OBE) testing effort was undertaken utilizing up to sixty DSRC radios which had the following objectives:

1. Analyze how well the communication channel operates, primarily in terms of Packet Error Rate (PER) and the Inter-

Packet Gap (IPG) distribution, in a variety of channel configurations and transmit characteristics.

2. Gain experience in the set-up and execution of a large-scale, DSRC test effort and in the areas of tools development, software tools, efficient logistics, setup, procedures, and analysis to ensure the end results are correct, meaningful, and repeatable.

The multiple OBE scalability testing conducted and the results obtained are described in this paper. Based on the results it is clear that using a dedicated, full-time, safety channel to transmit V2V safety messages provides superior performance over any of the other channel configuration methods employing IEEE 1609.4 channel switching when considering the PER and IPG metrics.

INTRODUCTION

The United States Department of Transportation (USDOT) and the Crash Avoidance Metrics Partnership-Vehicle Safety Communications 2 (CAMP-VSC2) Consortium (Ford, General Motors, Honda, Mercedes-Benz, and Toyota) initiated, in December 2006, a three-year collaborative effort in the area of wireless-based safety applications under the Vehicle Safety Communications- Applications (VSC-A) Project. One of the main achievements of the project is the implementation, testing, verification, and standardization of a safety message that supports all of the VSC-A safety applications. The result is the Basic Safety Message (BSM) as defined in the SAE J2735 Message Set Dictionary standard [1]. This message set is crucial in that it provides the foundation for the required Vehicle-to-Vehicle (V2V) safety system interoperability between automotive manufacturers.

Understanding how Dedicated Short Range Communications (DSRC) will perform as larger numbers of DSRC radios are added to the system (i.e., system scalability) is crucial for deployment of V2V safety systems. Following the successful completion of VSC-A Project objective testing activities [5], a preliminary multiple On-Board Equipment (OBE) scalability testing effort was undertaken utilizing up to 60 DSRC radios.

The primary objectives of the testing were to:

1. Gather the necessary data in order to analyze how well the communication channel operates, primarily in terms of Packet Error Rate (PER) and the Inter-Packet Gap (IPG) distribution, in a variety of channel configurations and transmit characteristics as well as a varying total number of radios

2. Gain experience in the set-up and execution of a large scale DSRC test effort and in the areas of tools development, software (SW) tools, efficient logistics, setup, procedures, and analysis to ensure the end results are correct and repeatable

Secondary goals and achievements confirmed that the operation of the VSC-A system, including the core and safety application modules, was not adversely affected during each of the scaling increments. Some of the scalability tests included on-board security, based on IEEE 1609.2 Elliptic Curve Digital Signature Algorithm (ECDSA) [2] and Verify-on-Demand (VoD) [4].

The OBE implementation [5] was enhanced to enable the emulation of two OBEs via dual-radio functionality. Self-contained DSRC enclosures (pods) were developed as a cost-effective approach for increasing the number of radios in the scalability test to the maximum achievable level of sixty units. In addition, to aid in ensuring testing was efficient, repeatable, and correct, a 2.4GHz Wi-Fi™ wireless mesh network, which enabled communication with each of the OBEs from a single point, along with scripts for command and control of the OBEs, were developed. Finally, the Over-the-Air (OTA) data was supplemented with a few additional data elements to ensure, in real-time, the ability to verify that the proper configuration was being used by all the radios during each test run.

Four channel configurations were defined for this testing. Three of the channel configurations utilized IEEE 1609.4 [3] channel switching and two of its variants. The fourth one utilized full-time access to Channel 172 (the safety channel).

The rest of this paper describes the multiple-OBE testing activities and provides an analysis of the results obtained based on the testing.

Note: Throughout this paper the term Host Vehicle (HV) refers to the OBE that was the focal point for the data collection and subsequent analysis for a particular test and the term Remote Vehicle (RV) refers to any other OBE, participating in the test, whose data was being collected by the HV. Each test had multiple HVs to allow the data from the test to be analyzed from multiple test setup perspectives, however, the data presented on any individual chart is from the perspective of a single HV.

DSRC POD DEVELOPMENT

In preparation for the multiple-OBE testing, the following activities were undertaken:

1. The OBE implementation, which supports dual-radio operation, was enhanced to enable the OBE to emulate two separate OBEs via the dual radios

2. A second OBE was integrated into the vehicle system test bed in such a way that full duplication of the system test bed hardware (HW) was not required by splitting Global Positioning System (GPS) and vehicle Controller Area Network (CAN) inputs. In combination with activity 1 above, this effectively brought the total number of radios per vehicle to four.

3. Self-contained DSRC pods running a GPS-only mode were developed to increase the total number of potential communication nodes to 60

Due to the limited number of OBEs that could be adequately installed in the vehicles and with the aid of the dual-radio functionality, the DSRC pod development was a critical and cost-effective means of increasing the number of radios to the desired level for the scalability testing efforts. The design of the pods allowed for them to be deployed in a stationary configuration via a tripod mount or in a mobile configuration via magnetic mount to the roof of a vehicle. The magnetic mounting capability, while not critical for the preliminary scalability testing effort, will potentially be more valuable in future scalability testing efforts. It allows the increase of the number of mobile DSRC-capable vehicles without having to fully equip vehicles and, thus, minimize costs. Figure 1 and Figure 2 below detail some of the internal and external design aspects of the DSRC pods which include:

- Internal Top Level: Steel Plate for mounting DSRC antenna, GPS antenna / receiver, and Wi-Fi antenna(s)
- Internal Middle Level: Holds OBE, 2.4GHz Wi-Fi Wireless Router, and optional equipment via side panel mounting
- Internal Bottom Level: Holds 12 Amp hour NiMH battery and steel box for internal component connection to external equipment
- External Enclosure: Made of material that is transparent to radio frequencies in the 5.9 GHz DSRC frequency band
- External Base Plate: Steel box for external equipment connection to internal components, magnetic mounts for attaching to the roof of a vehicle, 5/8 - 11 tripod mounting point

Figure 1. DSRC Pod Internal Structure Diagram

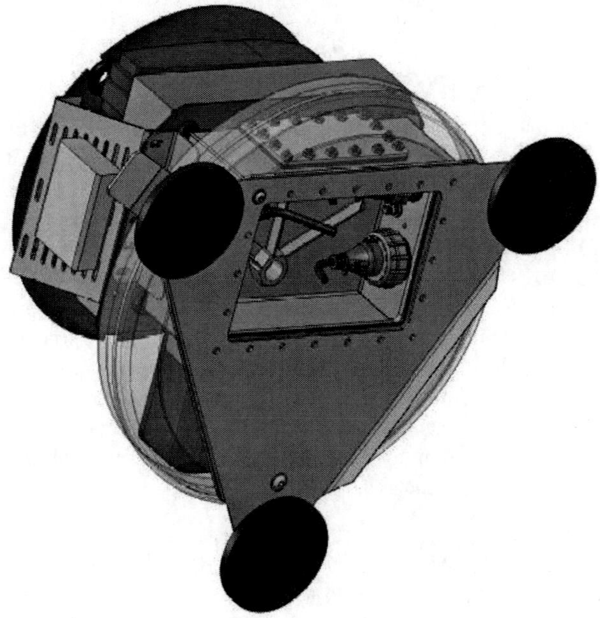

Figure 2. DSRC Pod Base Plate Diagram

TEST COMMAND, CONTROL, AND CONFIRMATION

Early in the project, a smaller scale, project testing effort exposed the difficulties in manually configuring and controlling multiple OBEs and ensuring the testing was correct, repeatable, and efficient. Therefore, the team made sure the preparation process included a number of steps to address these difficulties:

1. The use of a 2.4GHz Wi-Fi wireless network which enabled communication with each of the OBEs from a single point

2. The development of scripts for command and control of the OBEs via the 2.4GHz Wi-Fi wireless network

3. The addition of the applicable test number to the corresponding OTA message to ensure that early on in the field testing each OBE in the network was running the proper test configuration

WIRELESS NETWORK

A 2.4GHz Wi-Fi wireless router was installed and connected to the OBE(s) in each of the vehicles and pods. The routers were configured to utilize a light-weight mesh network protocol. Individual nodes use this topology information to compute the best path to a destination using and minimum number of relays or "hops." This provided a single central point for command and control of the entire system over a large area (see Figure 14 and Figure 15 in Appendix B for the vehicle and pod deployment configurations used during the testing) without requiring use of high power communications equipment.

SCRIPT DEVELOPMENT

With the 2.4GHz Wi-Fi wireless network in place, a set of scripts were developed to take advantage of the single central point of command and control that the network offered. For each test the scripts allowed for pushing the required configuration files to each of the OBEs, starting up each of the OBEs at the same time, and shutting down each of the OBEs at the same time at the end of the test. The scripts were also used to confirm that each test data log was appropriately captured on the OBEs upon completion of the test. The scripts, in combination with the network, allowed for confidence in test repeatability, efficient test execution, and flexibility in adding and/or modifying tests without requiring manual access to each OBE.

TEST NUMBER IDENTIFICATION

To confirm that the same test was being run on each of the OBEs for each test, the startup scripts provided a unique test number to the OBE at startup. This test number was then appended to all OTA transmissions from the OBE. A special Engineering Graphical User Interface (EGUI) screen was developed to display the test number of the test the HV was running as well highlight in red any of the RVs whose test number did not match that of the HV.

CHANNEL AND TEST CONFIGURATIONS

Four different channel configurations were identified along with seven tests exhibiting different transmit characteristics in order to determine how the channel behaved for each of the tested channel configuration/transmit characteristic combinations. This was one of the primary goals of the testing.

CHANNEL CONFIGURATIONS

Testing by the VSC-A team showed that, even with smaller numbers of radios, utilization of the IEEE 1609.4 channel switching mode without any changes leads to an increase in the observed PER due to "synchronized collisions." This is caused by the application layer being unaware of the start point and end point of the Control Channel (CCH) and/or the Service Channel (SCH) interval (defined by IEEE 1609.4). This increases the likelihood of the application layer attempting to transmit a message in a channel interval other than the intended one. If this occurs, the Media Access Control (MAC) layer will hold on to the message and transmit it at the beginning of the appropriate channel interval. Even with the back-off mechanism defined in IEEE 802.11, if enough radios are present in the system, the likelihood of having synchronized collisions increases. Table 1 below lists the channel configurations which were tested.

Table 1. Channel Configurations for Scalability Testing

Channel Configuration #	Channel Configuration Description
C1	IEEE 1609.4 channel switching mode
C2	Channel 172 dedicated safety channel (i.e., no channel switching)
C3	IEEE 1609.4 channel switching mode with messages submitted for transmission at a random time during each control channel interval
C4	IEEE 1609.4 channel switching mode with messages submitted for transmission via a time-shifting algorithm in an attempt to evenly space transmissions out during the intended channel

Configuration 1 (C1) was included in the scalability testing to confirm the synchronized collision issue as well as to serve as a baseline for the test results from the other channel configurations. Configurations 3 and 4 (C3 and C4) are countermeasures which were developed in an attempt to address the synchronized collision issue and in turn decrease the PER encountered when employing IEEE 1609.4 channel switching. Finally, Configuration 2 (C2) does not employ channel switching which provided full-time access to the channel essentially removing the artificial boundaries created by the CCH and SCH intervals as well as the need for the transmission guard interval at these boundaries. This more than doubles the bandwidth for transmission which decreases the likelihood of two radios transmitting at the same time, and, thus, was expected to decrease the PER observed over any of the other channel configurations tested.

TEST CONFIGURATIONS

Like the channel configurations, a baseline test configuration was defined. This configuration was tested for all four channel configurations in combination with all four DSRC radio scaling increments (24, 36, 48, and 60 radios). Some of the primary configuration settings for the baseline test configuration included:

- Stationary vehicles
- Transmit timer re-randomized every 30 seconds (not applicable for C3)
- Safety applications disabled
- Security disabled

• 222 bytes of extra padding added to OTA messages (total packet size 378 bytes)
• 10 Hz message transmit rate
• 6 Mbps data transmit rate

Note that the 222 bytes of extra padding was included to account for the security overhead that would have been present had security been enabled and a certificate attached to each message.

The baseline test configuration was the primary configuration used for analyzing channel behavior for each of the four channel configurations listed in Table 1. Seven additional tests were defined and were run on a sub-set of the channel configurations and scaling increments. These tests varied one or more of the baseline test configuration settings in order to analyze what effect these settings had on the channel configurations when compared to the results of the baseline test configuration. Table 2 in Appendix A provides a list of all the tests run along with the channel configurations and radio scaling increments that were tested for each test configuration. This paper provides the detailed analysis for the baseline stationary tests (Test # 1 in Table 2) and baseline moving tests (Test #2 in Table 2). For the other tests a brief discussion of the results is provided in the 'Summary/ Conclusions' section of this paper.

SAMPLE STATIONARY SCALABILITY TEST RESULTS

For the stationary test data analysis, gathering the necessary data in order to analyze the PER and the IPG distribution was of primary interest. Testing the baseline test configuration (Test # 1 in Table 2), in combination with each of the DSRC radio scaling increments for each of the four channel configurations listed in Table 1 above, was the initial focus of the testing activities. The results of this testing are presented in this section along with a brief discussion on some of the other non-baseline stationary test results.

Note that for any given radio scaling increment the number of expected receive signals for data analysis was two less than the number of transmit signals. This was due to the primary radio on an OBE not receiving it's own transmit signal as well as the primary radio on an OBE filtering out the signals it received from it's secondary radio. The filtering of the second radio was done in order to simplify the dual radio implementation.

TEST POD & VEHICLE DEPLOYMENT

As previously mentioned, each OBE was able to emulate two separate OBEs via the dual radio functionality supported by the OBE. There were eleven pods, each with one OBE, providing for 22 radios. In addition, there were 10 vehicles, 9 of which had 2 OBEs and 1 that had a single OBE which added an additional 38 radios for a grand total of 60 radios and thus 60 emulated OBEs.

The stationary pod/vehicle layout consisted of a center cluster of 4 pods and 5 vehicles with the remaining pods and vehicles placed at varying distances up to 275 m from the center cluster. Figure 14 in Appendix B provides a diagram identifying the location of each of the pods and vehicles used in the stationary tests.

The layout of the pods and vehicles was such that the center cluster OBEs were in communication range of all the other OBEs, whereas the OBEs furthest from the center cluster were not within communication range of one another. It should be noted that the primary radio on the OBE supported ~19-20 dBm output power, whereas the secondary radio supported ~14.5-16 dBm output power. It was the output power of the secondary radio that influenced selecting 275 m as the maximum range from the center cluster.

As previously mentioned, four DSRC radio scaling increments were tested consisting of 24, 36, 48, and 60 transmitting radios. In general the lower scaling increments included pods and vehicles closer to the center cluster while the ones that were further away were included as the scaling increments increased.

Note that due to the multi-tasking nature of the OBE, at the application layer, the OBE was not capable of providing both a non-emulated OBE packet and an emulated OBE packet to the radio HW of the OBE at precisely the same time. However, due to each of the OBE radios operating independently of one-another, once the packets were obtained by the radio HW, the physical (PHY) characteristics of the channel takes over. Thus, while it was unlikely that the 2 radios on the same OBE would provide contention with each other, it was expected that the 2 radios would be equally likely to provide contention with other remote OBEs in the test. Therefore, even though emulated OBEs were used in these tests via the dual-radio functionality of the OBE, the test results are expected to be closely representative to those obtained had non-emulated OBEs been used instead.

PACKET ERROR RATE TEST RESULTS

Figure 3, Figure 4, Figure 5 below show the results of PER analysis from the perspective of vehicle V2 as the HV which was part of the center cluster (see Figure 14 in Appendix B). C1 (1609.4-Timer Based), C3 (1609.4-Random Control Channel Interval Transmit), and C2 (Dedicated Safety Channel 172) results are shown for the 60 radio scaling increment tests only, however, the trends for all scaling increments were similar to the ones presented. The results from C4 (1609.4-Time Shifter) were very similar to C3 thus only the C3 results are presented.

The PER results are provided as bar graphs with the radios grouped into 4 PER bins of 0% - 10%, 10% - 30%, 30% - 50%, and 50% -100%. The PER bins are plotted along the horizontal axis with the number of radios falling within each PER bin plotted along the vertical axis.

The results show that the configuration method used for message transmission has a strong correlation to PER encountered. As expected, collisions at the beginning of a channel interval result in higher PER for C1 which has the worst performance. Taking advantage of knowing when the channel interval begins and ends, as well as implementing countermeasures in an attempt to avoid collisions as in C3 (and C4 which is not shown), provided better results than C1, which made no such attempt. C2, which provided full-time access to the channel, had the best performance and did not appear to be as affected as the other configurations as the scaling increments increased.

Worst Performance

- 60 Tx / 58 Rx radios total for test
- 2 radios not received
- 32% of radios with > 50% PER
- 71% of radios with > 30% PER
- 29% of radios with < 30% PER

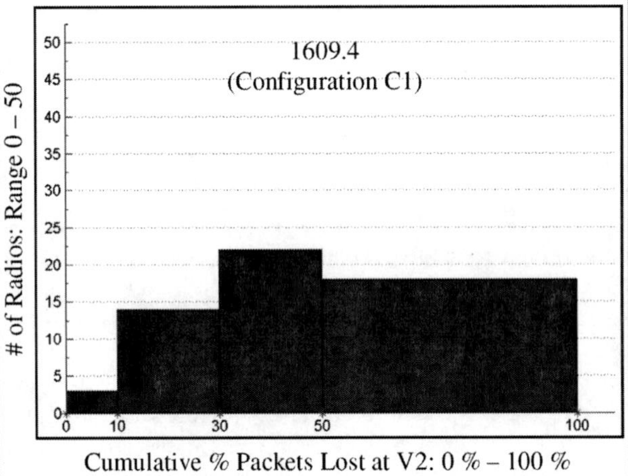

Figure 3. PER for Channel Configuration 1 w/ 60 Radios

Better Performance

- 60 Tx / 58 Rx radios total for test
- 1 radio not received
- 10% of radios with > 50% PER
- 39% of radios with > 30% PER
- 61% of radios with < 30% PER

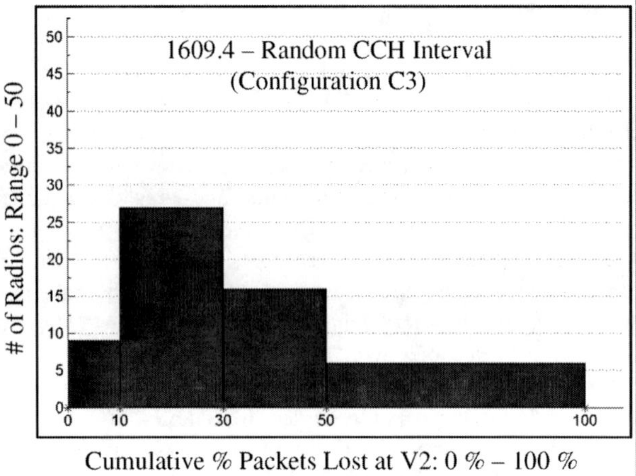

Figure 4. PER for Channel Configuration 3 w/ 60 Radios

Best Performance

- 60 Tx / 58 Rx radios total for test
- All radios received
- 100% of radios with < 30% PER

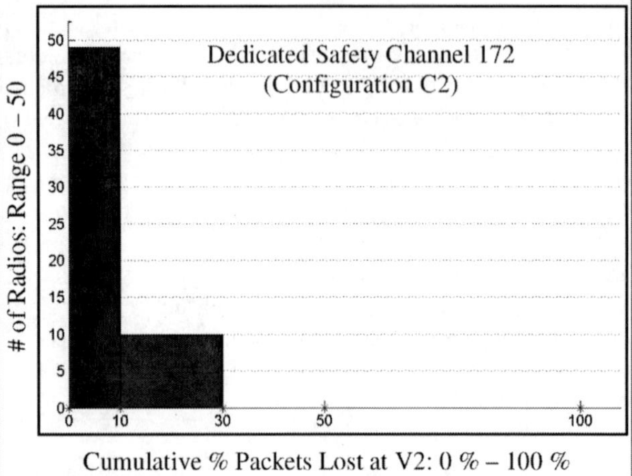

Figure 5. PER for Channel Configuration 2 w/ 60 Radios

INTER-PACKET GAP TEST RESULTS

The IPG results are provided as bar graphs with the radios grouped into 8 IPG bins of 0ms - 100ms, 100ms - 105ms, 105ms - 110ms, 110ms - 115ms, 115ms - 125ms, 125ms - 150ms, 150ms - 175ms, and 175ms - 200ms. The IPG bins are plotted along the horizontal axis with the number of radios falling within each IPG bin plotted along the vertical axis.

Figure 6, Figure 7, Figure 8 below show the results of the average IPG analysis from the perspective of vehicle V2 as the HV for the same channel configurations and radio scaling increment as the PER analysis. The trends for each of the channel configurations and scaling increment are similar to that of the PER results which is to be expected as an increase in PER should lead to an increase in the average time between receiving consecutive messages. Once again, C1 performed the worst, followed by C3 (and C4 which is not shown). C2 performed the best. What should be noted is that, the figures only include the radios whose average IPG was less than or equal to 200 ms. In taking this into consideration, for the 60 radio test C1 had only 38 of the possible 58 radios that met this condition, C3 had 51 of 58 radios that met this condition, and C2 had 57 of 58 radios that met this condition.

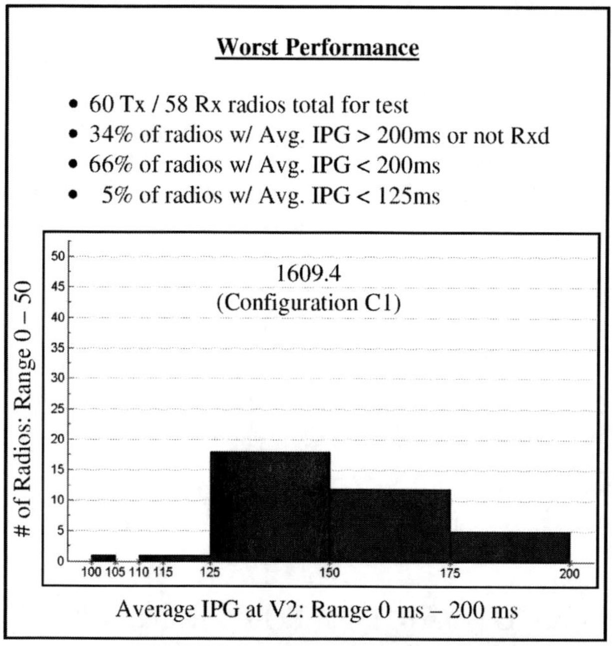

Figure 6. Avg. IPG for Channel Configuration 1 w/ 60 Radios

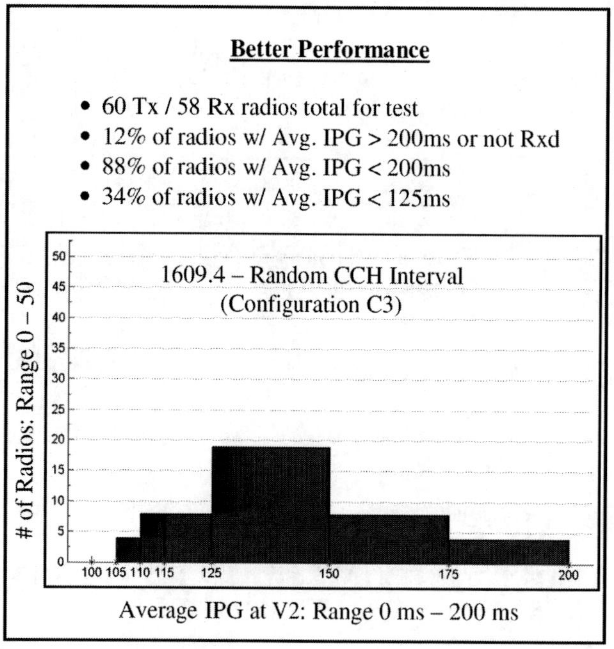

Figure 7. Avg. IPG for Channel Configuration 3 w/ 60 Radios

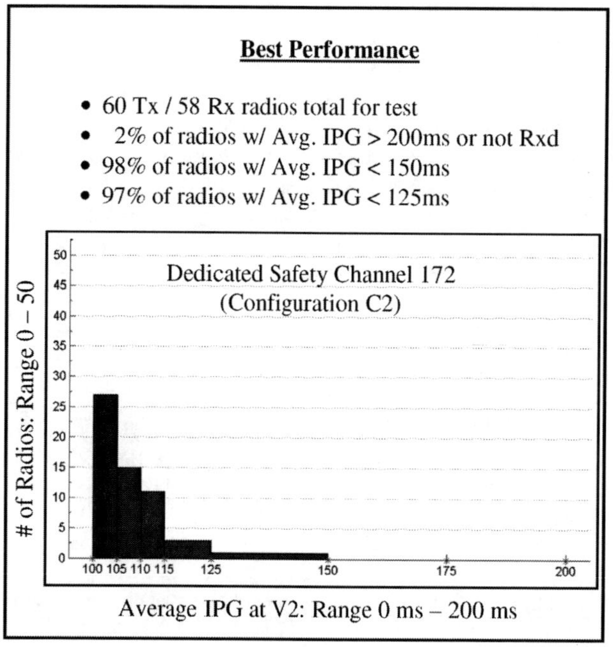

Figure 8. Avg. IPG for Channel Configuration 2 w/ 60 Radios

OTHER STATIONARY TESTS

In addition to the PER and IPG test results discussed above, other stationary baseline test analysis looked at the PER versus Range and the PER versus Received Signal Strength (RSS) for each of the configurations. For each of these tests, C2 outperformed C3 and C4, which in turn, outperformed C1 as in the previous test results.

The other non-baseline stationary tests primarily focused on C2 and C3 for the data gathering and analysis. The data analysis looked at the:

- PER versus a 5 Hz message transmit interval as opposed to the 10 Hz rate used in the baseline test
- PER versus 12 Mbps data rate as opposed to the 6 Mbps rate used in the baseline test
- PER versus increased packet length

SAMPLE MOVING SCALABILITY TEST RESULTS

In addition to the stationary deployment tests, a number of moving tests were run to analyze the effects of PER versus distance in a moving environment. The data analysis that follows for these moving tests (Test # 2 in Table 2) primarily looked at the PER versus distance from both an increasing range to the RV(s) and a decreasing range to the RV(s) from the HV's perspective. This included looking at the PER at the HV among all of the other RVs in the test in addition to the PER at the HV with respect to the single RV that the HV was traveling with during the test. Channel configurations C2 and C3 were tested for three radio scaling increments consisting of 24, 48, and 60 radios. Other than having moving vehicles, the test configuration was the same as the baseline test configuration. Only the data from the 60 radio scaling increment are presented.

OBE MOVING TEST POD & VEHICLE DEPLOYMENT

For consistency, the pod / vehicle layout for the pods and vehicles that remained stationary did not change considerably from the all-stationary tests. It consisted of a center cluster of four pods and four vehicles with the remaining pods placed at varying distances up to 275m from the center cluster. Unlike the stationary tests all vehicles outside of the center cluster were moving for these tests.

Figure 15 in Appendix B provides a diagram identifying the location of each of the pods and vehicles that were used in the tests and identifies which vehicles were stationary and which were moving. For the moving vehicles, vehicles V2 (HV) and V1 (RV) traveled together with V2 following relatively close behind V1. Similarly vehicles V3 (HV) and V4 (RV) traveled together with V3 following relatively close behind V4. All four vehicles traveled in a big loop thru the main track. Vehicles V6 (HV) and V5 (RV) made a smaller loop outside of the track with each one attempting to remain at opposite ends of the loop.

INTERPRETING THE CHARTS

The following data analysis sections start with a comparison between channel configurations C2 and C3 to show that, similar to the stationary tests, C2 performs better than C3 from a PER analysis perspective. The remaining data analysis only provides the data for channel configuration C2. Since the data is similar for vehicles V2 and V3, which were both traveling in a big loop through the main track, only the data from V2 will be presented in order to allow for a comparison between moving test results and the stationary test results which are also from the perspective of V2.

Figure 16 in Appendix C shows the charts that were developed to analyze the PER versus range from the HV perspective for all of the RVs the HV was in communication with. Additionally, similar charts are also presented in the analysis sections that show the PER versus range from the HV perspective for the RV that the HV was traveling with. To aid in the plotting of the data, the ranges were grouped into 3m bins covering a range of 0 to 500m. Two types of charts were developed:

- A chart to plot the number of packets received at each range grouping (Range: 0 to 4000 packets)
- A chart to plot the percentage of packets lost or PER for each range grouping (Range: 0 to 100%)

Each of these charts has multiple plots:

- Blue lines / dots show the # Packets / PER for all of the RVs
- Red lines / arrows show the # Packets / PER when the HV to RV distance was decreasing
- Green lines / arrows show the # Packets / PER when the HV to RV distance was increasing or unchanged

For the PER charts linear (solid line) and quadratic fit (dashed line) curves are provided based on the plotted dots or arrows.

PER COMPARISON FOR CHANNEL CONFIGURATION C2 VERSUS C3

Figure 9 and Figure 10 show a comparison of the results between channel configuration C2 and channel configuration C3 from the perspective of vehicle V2 as the HV. Similar to the stationary tests, C2 has better PER versus range performance, for all ranges both increasing and decreasing, than C3.

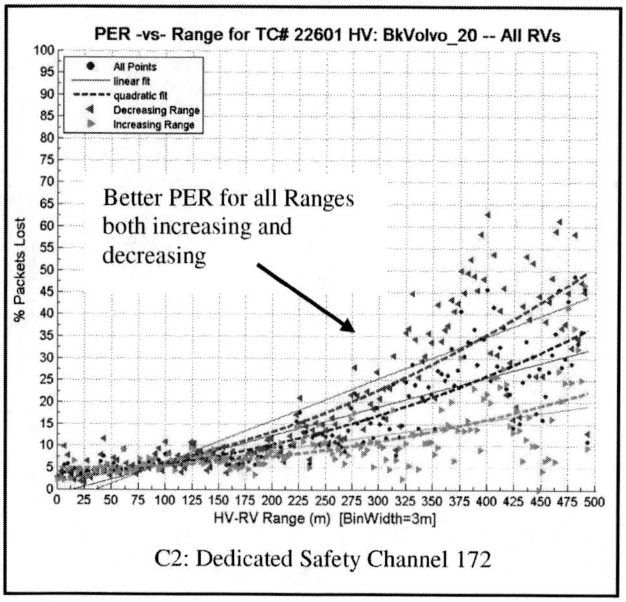

Figure 9. PER vs. Range for Moving Vehicles - C2 w/ 60 Radios

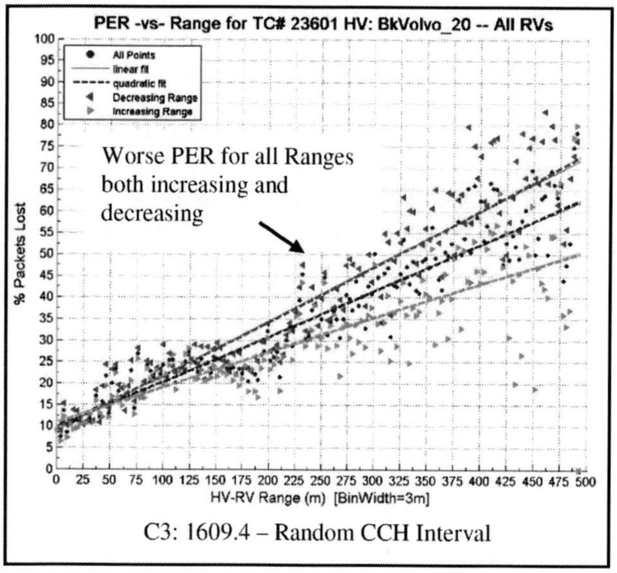

Figure 10. PER vs. Range for Moving Vehicles - C3 w/ 60 Radios

CUMULATIVE PER FOR MOVING HV WITH MOVING BLOCKING RV

Figure 11 shows the PER versus range for a moving HV (V2) with a moving blocking RV (V1) for all of the RVs the HV was in communication with. The top chart in the figure shows that packets were received from other vehicles at all distances from 0m to 500m, but most vehicles were within 250m due to the test layout and driving patterns as depicted in Figure 15 in Appendix B. The bottom chart in the figure shows that the PER from RVs located in front of the HV (decreasing range) is worse than from RVs located behind (increasing range). This difference is more noticeable at greater distances. This may be caused from the RV being located in front of the HV, thereby reducing the ability for the HV to receive messages from the forward direction.

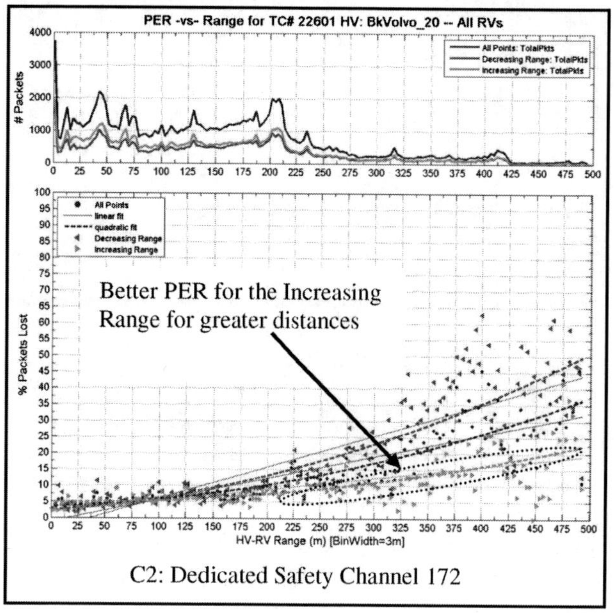

Figure 11. PER vs. Range for Multiple RVs - Moving HV w/ Blocking RV - C2 w/ 60 Radios

Figure 12 shows the PER versus range from the HV perspective for the moving RV that the HV was traveling behind. The top chart shows the distance between the HV and RV ranged between 10m to 60m while the bottom chart shows that the PER from the leading RV to the following HV was less than 10% for most of the distances measured. The congestion level of 60 transmitting radios did not appear to affect the PER of the RV at these relatively close distances.

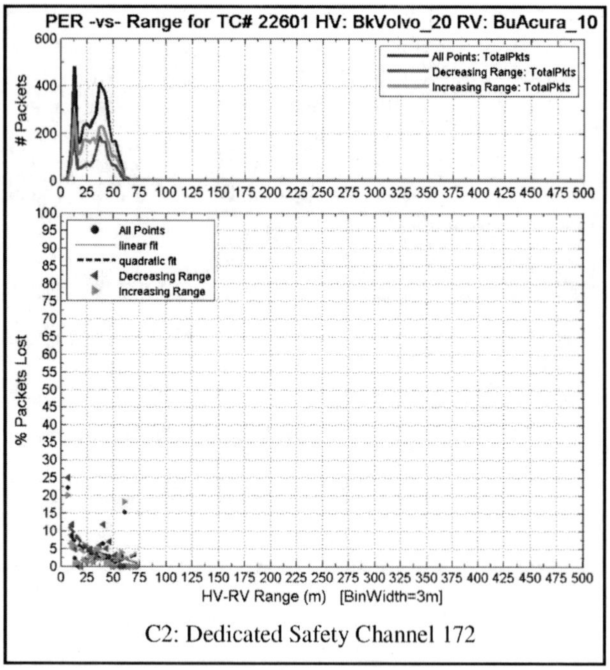

Figure 12. PER vs. Range for Principle RV - Moving HV w/ Blocking RV - C2 w/ 60 Radios

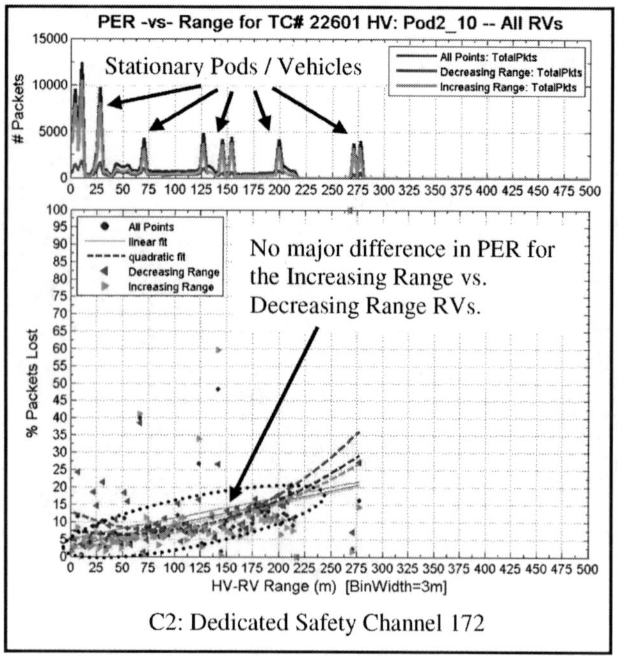

Figure 13. PER vs. Range for Multiple RVs - Stationary HV - C2 w/ 60 Radios

CUMULATIVE PER FOR STATIONARY HV

Figure 13 shows the PER versus range for a stationary HV (Pod 2) for all of the RVs the HV was in communication with. The top chart shows that most packets received by the HV were at specific distances. Since the HV was stationary these correspond to the stationary RVs (pods and vehicles) in the test. The packets from the moving RVs were received at distances from 0-225m with the furthest stationary pod being at 275m. Recall that packets received from an RV where there is no change in the distance are categorized as "Increasing Range," thus, the green spikes for each of RVs that are stationary with respect to the HV. The bottom chart shows that the PER from all RVs moving towards or away from the stationary HV appears similar. Additional PER results (not shown) between Pod2 and specific RVs also do not show any clear difference in either direction.

SUMMARY/CONCLUSIONS

In general, the stationary test results show that using a dedicated, full-time, safety channel to transmit V2V safety messages clearly provides superior performance over any of the other channel configuration methods tested employing IEEE 1609.4 channel switching when considering the following PER and IPG metrics:

- Cumulative PER
- PER compared to RSS
- PER compared to Range
- Average IPG
- Worse Case IPG

Results from other stationary tests that were run show that:

1. Transmitting smaller packets has a positive effect on the PER

2. Decreasing the message transmit rate (e.g., 5 Hz versus 10 Hz) has a positive effect on PER

3. Increasing the data transmit rate (e.g., 12 Mbps versus 6 Mbps) limits the number of RVs in communication range; however, it appears to have a positive effect on the PER for RVs with stronger signals

For the moving tests, like the stationary test results, the dedicated safety channel configuration (C2) results in superior performance, when considering PER, compared to using the CCH interval (C3) for transmitting periodic safety messages.

Additionally the moving test results show:

1. A greater range of best case and worse case PER. While not conclusive, the difference in PER appears to be caused from blockage from other vehicles, both moving and stationary.

2. The PER between adjacent moving vehicles (less than 60m apart) was less than 10% with 60 RVs when using the dedicated safety channel (C2)

3. That while the stationary PER test results were overall better than the moving test results, but the "better case" moving PER test results (i.e., presumed without blockage) were similar to the stationary results

These results are a good start for beginning to understand the effects that combinations of stationary only and moving in combination with stationary vehicles may have on PER at a particular HV. For the moving scenarios more analysis needs to be done on the affect a blocking vehicle may have on the PER at a particular HV as well as combinations of blocking vehicles (e.g., multiple vehicles blocking the HV, vehicle blocking the RV, etc.).

Some of the next steps include incorporating lessons learned into future projects where V2V system scalability has to be proven beyond the achievable total number of units used within the CAMP VSC2 VSC-A Project and presented in this paper (i.e., 60). This includes lessons learned in test bed design and development, SW design and stability, and scalability testing logistics.

REFERENCES

1. SAE International Surface Vehicle Standard, "Dedicated Short Range Communications (DSRC) Message Set Dictionary," SAE Standard J2735, Rev. Nov. 2009.

2. Intelligent Transportation Systems Committee of the IEEE Vehicular Society, "IEEE Trial-Use Standard for Wireless Access in Vehicular Environments - Security Services for Applications and Management Messages," IEEE Std 1609.2™, 2006

3. Intelligent Transportation Systems Committee of the IEEE Vehicular Society, "Draft Standard for Wireless Access in Vehicular Environments (WAVE) - Multi-channel Operation," IEEE P1609.4, Rev. D07, Dec. 2005

4. Krishnan, H., Technical Disclosure, "Verify-on-Demand - A Practical and Scalable Approach for Broadcast Authentication in Vehicle Safety Communication," IP.com number: IPCOM000175512D, IP.com Electronic Publication: October 10, 2008.

5. Ahmed-Zaid, F., Krishnan, H., Maile, M., Caminiti, L. et al., "Vehicle Safety Communications - Applications: System Design & Objective Testing Results," *SAE Int. J. Passeng. Cars – Mech. Syst.* **4**(1):417-434, 2011, doi: 10.4271/2011-01-0575.

CONTACT INFORMATION

Farid Ahmed-Zaid
Ford Motor Company
fahmedza@ford.com

ACKNOWLEDGMENTS

The CAMP VSC2 Participants would like to acknowledge the following USDOT personnel for their invaluable project support; Art Carter, Ray Resendes, and Mike Schagrin. The Participants would also like to thank Aaron Weinfield, Sue Graham, and Roger Berg from DENSO International America, Inc. - North America Research Laboratory for their technical project support.

DEFINITIONS/ABBREVIATIONS

BSM
Basic Safety Message

CAMP
Crash Avoidance Metrics Partnership

CAN
Controller Area Network

CCH
Control Channel

DSRC
Dedicated Short Range Communications

ECDSA
Elliptic Curve Digital Signature Algorithm

EGUI
Engineering Graphical User Interface

GPS
Global Positioning System

HV
Host Vehicle

HW
Hardware

IPG
Inter-Packet Gap

MAC
Media Access Control

OBE
On-Board Equipment

OTA
Over-the-Air

PER
Packet Error Rate

PHY
Physical (layer)

RSS
Received Signal Strength

RTCM
Radio Technical Commission for Maritime Services

RTK
Real-Time Kinematic

RV
Remote Vehicle

SCH
Service Channel

SW
Software

V2V
Vehicle-to-Vehicle

VoD
Verify-on-Demand

VSC2
Vehicle Safety Communications 2 (Consortium)

VSC-A
Vehicle Safety Communications -Applications

USDOT
United States Department of Transportation

DISCLAIMER

This material is based upon work supported by the National Highway Traffic Safety Administration under Cooperative Agreement No. DTNH22-05-H-01277. Any opinions, findings, and conclusions or recommendations expressed in this publication are those of the Author(s) and do not necessarily reflect the view of the National Highway Traffic Safety Administration

APPENDIX A

TEST CONFIGURATIONS FOR SCALABILITY TESTING

Table 2. Test Configurations for Scalability Testing

Test #	Channel Configuration Used				Scaling Increment: Number of Radios				Changes to the Baseline Test Configuration
	C1	C2	C3	C4	24	36	48	60	
1	X	X			X	X	X	X	• Baseline Test (See list above)
			X	X	X		X	X	
2		X	X		X		X	X	• Moving Vehicles
3		X	X		X		X	X	• 12 Mbps Data Transmit Rate
4.1		X	X		X	X	X	X	• Security Enabled and configured for ECDSA VoD • No extra padding added to OTA messages
4.2		X	X				X	X	• Moving Vehicles
									• Safety Applications Enabled • Security Enabled and configured for ECDSA VoD • No extra padding added to OTA messages
5		X	X		X		X	X	• 86 additional bytes of padding added to OTA message (to account for Radio Technical Commission for Maritime Services (RTCM) 1002 data that would be present if Real-Time Kinematic (RTK) positioning was enabled and seven satellites were in view)
6		X		X	X		X	X	• No Transmit timer re-randomization
0		X	X		X		X	X	• 5 Hz Message Transmit Rate

APPENDIX B

OBE DEPLOYMENT CONFIGURATIONS

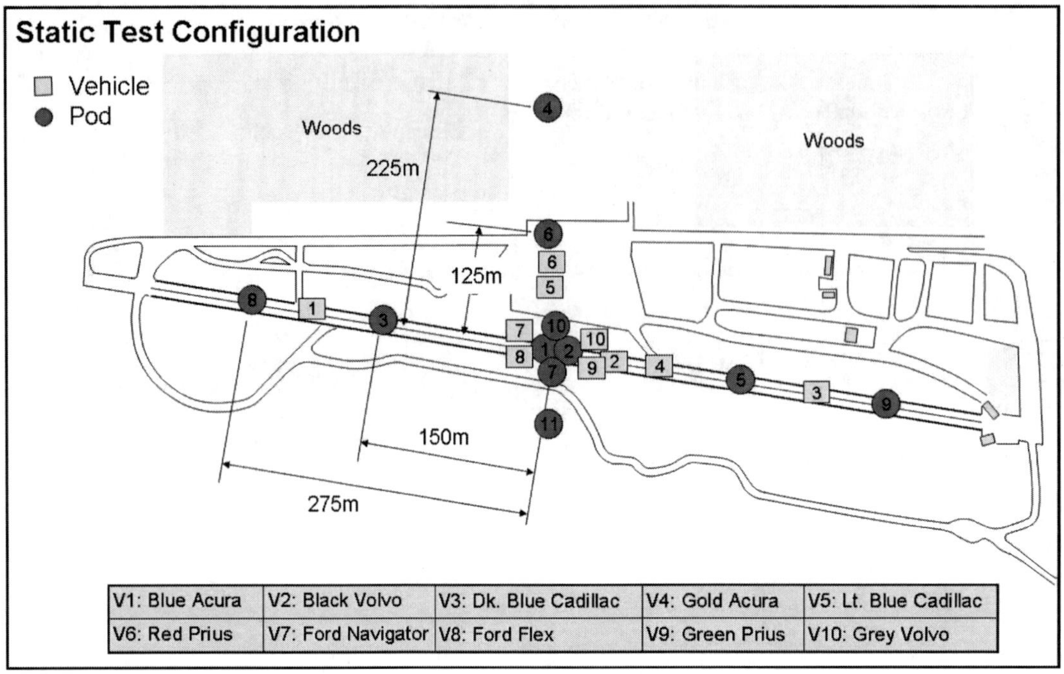

Figure 14. Vehicle and Pod Stationary Deployment Configuration

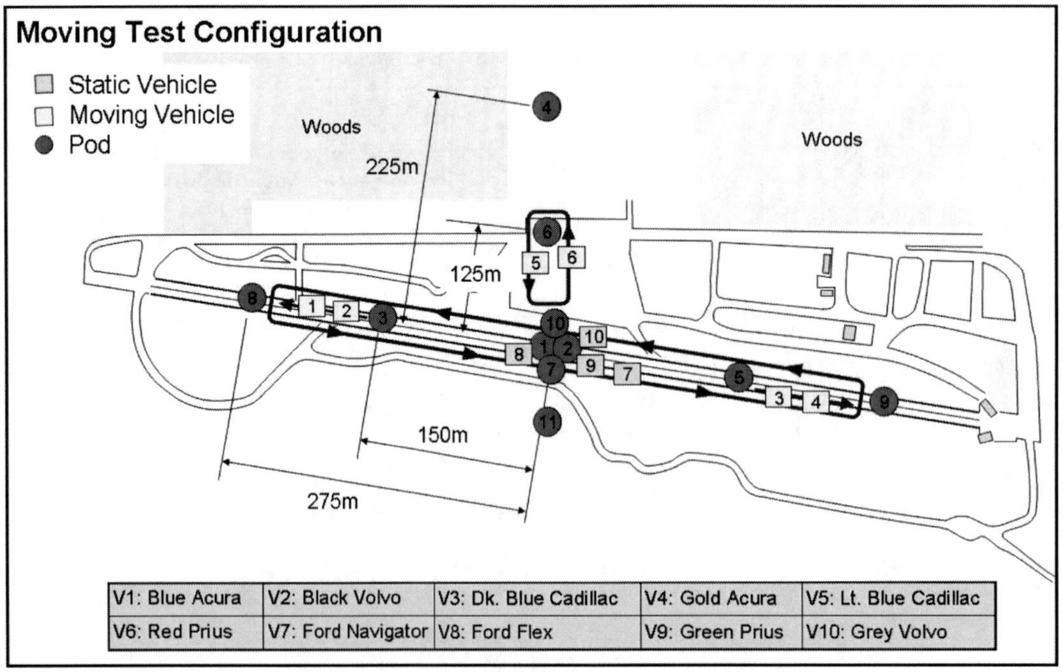

Figure 15. Vehicle and Pod Moving Deployment Configuration

APPENDIX C

INTERPRETING THE MOVING TEST CHART DATA

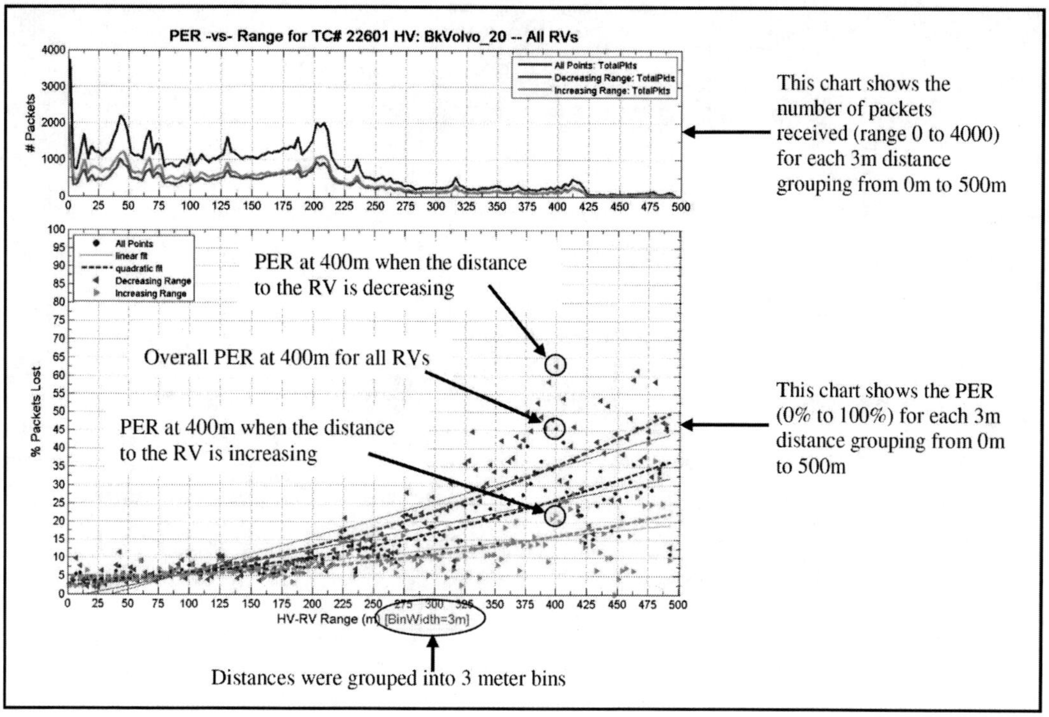

Figure 16. PER vs. Range for Moving Vehicles - Example Plot for 60 Tx Radios

SAE International

Understanding Driver Perceptions of a Vehicle to Vehicle (V2V) Communication System Using a Test Track Demonstration

2011-01-0577
Published
04/12/2011

Christopher Edwards and Jon Hankey
Virginia Tech Transportation Institute

Raymond Kiefer, Donald Grimm and Nina Leask
General Motors Company

doi:10.4271/2011-01-0577

ABSTRACT

Vehicle-to-vehicle (V2V) communication systems can enable a number of wireless-based vehicle features that can improve traffic safety, driver convenience, roadway efficiency, and facilitate many types of in-vehicle services. These systems have an extended communication range that can provide drivers with information about the position and movements of nearby V2Vequipped vehicles. Using this technology, these vehicles are able to communicate roadway events that are beyond the driver's view and provide advisory information that will aid drivers in avoiding collisions or congestion ahead. Given a typical communication range of 300 meters, drivers can potentially receive information well in advance of their arrival to a particular location. The timing and nature of presenting V2V information to the driver will vary depending on the nature and criticality of the scenario. The purpose of this study was to gather data on driver perceptions and opinions with respect to a V2V communication system. The 125 participants tested on the Virginia Smart Road experienced a V2V system capable of demonstrating various V2V applications. These applications included alerting the test participants to the following events ahead of the vehicle: vehicle braking hard, slow moving vehicle, vehicle with hazard warning lights activated, post-crash notification, electronic stability control (ESC) activation, and a potential intersection crash situation. The results indicated that the V2V applications demonstrated were well-received by test participants and were believed to provide safety benefits. These results, coupled with the relative affordability of V2V compared to other autonomous sensing systems, suggest that the V2V system may with anticipated future levels of deployment provide a promising approach to improve traffic safety and increase the penetration of safety systems across the vehicle fleet.

INTRODUCTION

Vehicle to Vehicle (V2V) wireless communication utilizes wireless technology to send information between vehicles that are similarly equipped. The vehicles can communicate and exchange information within a specific radius (e.g., ¼ mile) using global positioning system (GPS) capability and technology similar to Wi-Fi. For example, if a V2V-equipped vehicle brakes suddenly, this event can be relayed back to a following vehicle similarly equipped with V2V hardware which can trigger an alert (e.g., a flashing display and a beeping warning) to the following driver. More generally, V2V vehicles can exchange a wide range of safety, as well as non-safety, related information. The timing and nature of presenting information to the driver will vary depending on the nature and criticality of the scenario.

The purpose of this study was to gather data and information on driver perceptions and opinions with respect to potential near-term V2V applications. The prototype V2V system used a Driver Vehicle Interface (DVI) to communicate information associated with these applications to the driver.

Figure 1. V2V Vehicles and DVI display mounted above the center stack

METHOD

APPARATUS

The study used a combination of test vehicles that included two 2004 and two 2007 General Motors Cadillac CTS's equipped with prototype V2V technology, as well as two General Motors Cadillac sedans that were not equipped with V2V technology. The latter vehicles were driven by experimenters, and of the four V2V-equipped vehicles, two were driven by test participants and two were driven by experimenters.

A total of three vehicles constituted one experimental vehicle platoon (or research entity). For each platoon, a non-equipped V2V vehicle drove between the two equipped V2V vehicles. The lead (V2V-equipped) vehicle within a platoon, referred to as the "scenario vehicle", was driven by an experimenter responsible for setting up each scenario. The second vehicle within a platoon, referred to as the "confederate vehicle", was a non-equipped V2V vehicle driven by an experimenter. The last (V2V-equipped) vehicle within a platoon, referred to as the "participant vehicle", was driven by a test participant and included an additional one or two passenger test participants. These "passenger" participants rode in the backseat, with a clear view of the DVI display area, while the experimenter in the participant vehicle rode as a front seat passenger.

Figure 1 shows the V2V vehicles and position of the top-of-dashboard DVI. When a V2V event was encountered an icon was displayed on the left side of the DVI display simultaneously with an auditory alert (i.e., a series of five warning beeps). A total of four icons were used which were based on previous icon comprehension/rating research conducted by General Motors. These icons were chosen based on the V2V applications examined under this effort. Only one icon and corresponding auditory alert was issued for each of the six scenarios included in the study. It should be noted that one of the icons was employed for three different scenario types.

The V2V auditory and visual alerts in the participant vehicle were activated by the V2V system in four of the scenarios and manually triggered by the experimenter (via a "Wizard of Oz" approach) in two of the scenarios for experimental convenience purposes. In the latter case, at a key point during the scenario, the experimenter in the lead V2 V-equipped vehicle would press a button that "artificially" triggered the V2V alert in the following participant vehicle.

The vehicles were instrumented to obtain both audio and video recordings. Audio recordings were made using a microphone located near the driver's steering wheel. Video recordings were gathered using four cameras mounted within the cabin of the vehicle. These cameras provided the

Figure 2. V2V scenarios and corresponding icons that included: (1) a potential intersection crash situation, (2) vehicle braking hard, (3) ESC activation, (4) post-crash notification, (5) slow moving vehicle, and (6) vehicle with hazard warning lights activated.

following views: driver face, forward roadway, over-the-shoulder steering wheel and the V2V display.

Scenarios

A total of six V2V scenarios were examined which were linked to the corresponding icon presentations shown in Figure 2. These scenarios illustrated to the participants how V2V technology could alert them to various types of events in their direction of travel, as well as the V2V alerting approach associated with each event type. Each of these scenarios corresponds to a potential V2V application (or feature). The scenarios included:

1. A potential intersection collision situation where the scenario vehicle was positioned in the crossing path at an intersection. Based on predefined timing, the scenario vehicle would accelerate as if it was going to run across the intersection and run the red light, but then stop short of entering the intersection. The alert was presented when the participant vehicle was approximately 4 seconds from the intersection.

2. A hard braking scenario where the lead platoon vehicle initiated a “last second” right turn and rapidly decelerated. When the deceleration exceeded 0.3 g's, the hard braking ahead alert in the participant vehicle was activated.

3. The Electronic Stability Control (ESC) alert scenario involved a tire tread being placed in the roadway well in advance of the vehicle platoon. The ESC alert was activated in the participant vehicle when the scenario vehicle driver swerved aggressively to avoid the tire.

4. For the Post Crash Notification (PCN) scenario, the scenario vehicle accelerated well ahead of the other vehicles to the predetermined PCN location, and staged what appeared to be a minor collision situation with a roadside work zone barrel. An alert was issued to the V2V platoon vehicle up to ¼ mile before the mock accident site.

5. The slow moving vehicle scenario involved the scenario vehicle straddling the shoulder line with the right turn signal activated. The slow moving vehicle alert was received by the participant vehicle when the vehicle was within a ¼ mile radius of the scenario vehicle.

6. In the vehicle hazard light activated scenario, the scenario vehicle accelerated well ahead of the other vehicles to a location and pulled off to the shoulder of the roadway. Once positioned out of the participants' view (around a curve), the scenario vehicle driver placed a tire tread near the rear right tire of the vehicle (as if they were changing a tire) and activated the hazard warning lights.

The Virginia Smart Road

The Virginia Smart Road is a unique, state-of-the-art, full-scale, 2.2 mile closed test-bed research facility managed by the Virginia Tech Transportation Institute (VTTI) and owned and maintained by Virginia Department of Transportation (VDOT). The scenario event locations and the platoon stopping points along the test roadway are shown in Figure 3.

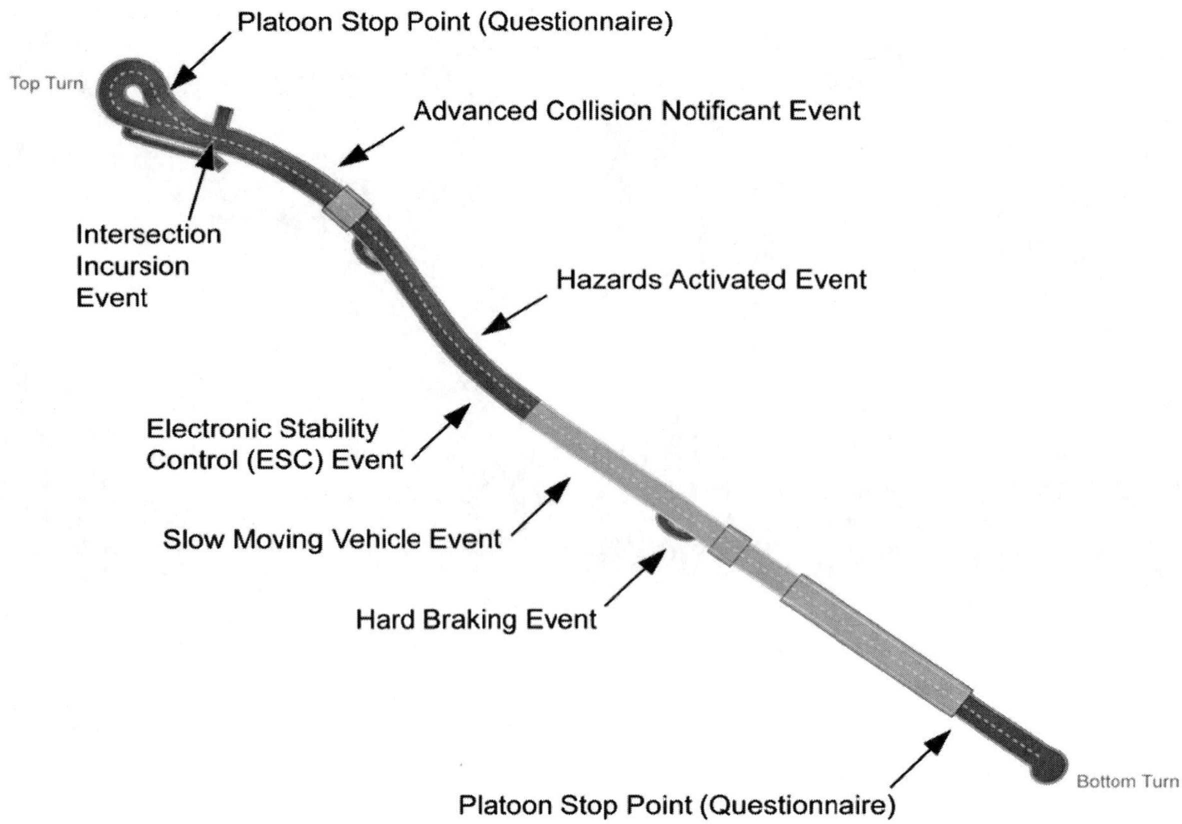

Figure 3. Virginia Smart Road Test Track and Scenario Locations. (*Note: ESC, hazards activated, and slow moving vehicle events were randomized between the three locations marked by these events in the figure).

These locations were chosen to enhance the realism of the various V2V scenarios. The intersection incursion, PCN, and hard braking ahead scenarios always occurred at the same pre-defined location. The remaining three scenarios (ESC, slow moving vehicle ahead, vehicle ahead with hazard warning lights activated) occurred in either an uphill or downhill direction on a relatively straight section at the midpoint of the test track. Figure 3 shows the locations of the scenario events. The ESC, slow moving vehicle, and hazard warning events were randomized between the three locations indicated by these events in the figure.

The stop points at each end of the roadway allowed the lead scenario vehicle to rejoin their respective vehicle platoon and clear the roadway for the next vehicle platoon to proceed. In addition, these stop points allowed the test participants to fill out a post-experience questionnaire about the scenario and their impression about the corresponding alert (DVI) approach that they had just experienced. The platoon vehicles were positioned such that one vehicle platoon could not view the scenarios the other vehicle platoon experienced.

PARTICIPANTS

Of the 125 participants, 63 were male and 62 were female. The average age of the entire sample was 51 years old. The age distribution of these participants is provided in Table 1.

Table 1. Age Groups Sample Sizes

Age Groups	Sample Size
18-29	20
30-39	16
40-49	13
50-59	29
60-69	33
70+	14
Total	125

Participants were paid $20 per hour, and were free to withdraw from the study at any time (none elected to withdraw). The typical test session, which included pre-drive paperwork, test drives, and post-drive questionnaires, lasted about three and a half hours.

STUDY DESIGN

The focus of this study was to obtain subjective feedback via questionnaires from the driver and passenger test participants riding in the participant vehicle. The questionnaires administered included a background (screening) questionnaire, a pre-drive questionnaire, a series of post-scenario questionnaires administered following each individual V2V scenario, and a post-drive questionnaire. Two-thirds of the participants were tested under daylight conditions (n=82), with remaining participants tested under nighttime conditions (n=43).

PROCEDURE

Upon arrival, participants were asked to sign an informed consent statement. After consenting to participate, a simple visual and auditory test was administered. Following successful visual and auditory screening, each participant completed a pre-drive questionnaire. The test participants were then randomly chosen to ride in one of the two V2V-equipped participant vehicles. Test participants were provided a brief overview of V2V technology prior to the feature demonstration drives, but were not provided information in advance on the specific nature of the V2V features prior to experiencing them during the demonstration. This latter experimental strategy was employed to determine the extent to which the alert approach provided an intuitive explanation of the scenario, and to loosely approximate what drivers might experience when first exposed to the various V2V technology features examined in this effort.

Participants within each participant vehicle decided among themselves who would drive the participant vehicle for the entire test. If more than one participant wanted to drive, the driver was chosen by a coin flip. After escorting the participants to the participant vehicles and verifying that the rear seat test participant passengers were comfortable, the in-vehicle experimenter introduced the vehicle to the participant driver. These drivers were shown how to adjust the mirrors, activate the windshield wipers, and adjust the driver's seat. The in-vehicle experimenter then reviewed the protocol for the driving portion of the study. In addition to outlining the speed limit (45 MPH) and explaining the practice lap portion of the study, the in-vehicle experimenter reviewed the V2V system, pointed out the V2V display, and confirmed the participant could see the V2V display.

Participants were then informed that they would experience a series of scenarios while riding in the test vehicle, and were instructed to maintain view of the V2V display during the V2V feature demonstration drives. Passenger participants were instructed to evaluate the scenario as if they were driving the vehicle. Participants were asked to refrain from discussing any comments or thoughts they had with others in the vehicle during their entire set of feature demonstration drives. In addition, participants were told that they would have the opportunity to openly share their views about their experiences with other participants and researcher after completing the entire test drive in a focus group session.

When the drivers and vehicles were ready to continue, each participant vehicle joined their respective vehicle platoon that included the confederate vehicle and the scenario vehicle. When each of the two vehicle platoons was ready to begin they drove to the Virginia Smart Road. On the test road, participants first experienced a practice lap to get familiar with the vehicle and the roadway. After this lap, participants experienced a different scenario during each "run" (or length) of the test road in a randomized order. All participants completed six runs and experienced each of the six V2V feature demonstration scenarios once. The test runs were choreographed such that the two vehicle platoon paths never crossed, and that both platoons experienced different orders of the six scenarios in order to minimize order bias.

After each V2V scenario, the participant vehicle drove to a pre-determined location (or platoon stop point) located at the top or bottom of the test track (see Figure 3). With the vehicle parked, participants were asked to individually complete a one-page written questionnaire concerning the V2V display/ feature they had just experienced. It is important to note that participants were not provided any advance information on the V2V demonstration scenario prior to experiencing the scenario. The in-vehicle questionnaire administered following each scenario contained a picture of the corresponding icon (shown on the V2V display) they had experienced during the scenario. The questionnaire asked participants to rate the timing of the alert, interpret the message/icon in their own words for each scenario, and rate the icon clarity, usefulness of the alert, perceived safety and willingness to have V2V on their next vehicle.

When each participant had completed the questionnaire for a given scenario, the in-vehicle experimenter gathered the questionnaires in order to prevent participants from looking back to their previous questionnaire responses.

After completion of the in-vehicle questionnaire for a given scenario, the in-vehicle experimenter confirmed that the road was clear for the next scenario and signaled the driver of the scenario vehicle to begin the next run. When all six scenarios had been completed, the platoon then exited the test road and returned to the research building. Participants then completed a post-drive questionnaire on the V2V system.

RESULTS

The result section discusses the participants' impressions of the V2V system. In cases where questions employed Likert scales (described below), Chi-Square statistics were conducted to examine the effects of gender (males versus females) and time of day (daytime versus nighttime). These

Table 2. Percentage breakdown of responses to the post scenario questions for the intersection collision danger event.

Question	Gender		Daytime/Nighttime	
	Male	Female	Day	Night
In my opinion, the message appeared: - At the right time (Too Late)	69% (31%)	52% (45%)	70%* (27%)	40%* (60%)
The icon clearly communicates the intended message? – Agree (Disagree)	71% (15%)	80% (12%)	74% (10%)	78% (20%)
The feature provides useful information? - Agree (Disagree)	84% (12%)	89% (8%)	89% (8%)	82% (16%)
The feature would increase my driving safety? - Agree (Disagree)	79% (13%)	86% (10%)	85% (8%)	76% (16%)
I want this feature on my next vehicle? Agree (Disagree)	75% (17%)	83% (10%)	82% (10%)	75% (21%)

(Significant difference between groups $p < 0.05$)*

paired comparisons were made by collapsing the agree/disagree categories and removing the neutral category. Unless indicated, no significant differences were found between group pairs ($p < 0.05$ criterion was used as the statistical significance criterion).

For each of the V2V scenarios, participants were asked to provide feedback on a variety of elements. First, participants were asked to describe in their own words what the DVI icon was communicating during the scenario and rate the appropriateness of the alert timing ("too early", "at the right time", or "too late"). The participant interpretations were categorized based on commonalities. Participants were then asked to rate four statements on a 5-point Likert scale (where: 1= Strongly Disagree, 2 = Disagree, 3 = Neutral, 4= Agree and 5= Strongly Agree). These statements addressed the extent to which the alert communicated the intended message (i.e., message interpretability), the alert was useful, the alert would increase their driving safety, and the alert was desirable in their next vehicle.

INTERSECTION COLLISION DANGER

As shown in Table 2, 60% of the participants (ranging between 40% - 70% across groups) rated the timing of the intersection collision danger alert as occurring at the right time. A significant difference was found between daytime and nighttime groups, $\chi2(1, N=119) = 11.58$, $p < 0.01$, with the nighttime group rating the alert as occurring "too late" in 60% of the cases compared to 27% under daytime conditions. This finding may be due to the added difficulty to observe this scenario under nighttime conditions. The remaining questions addressing icon usefulness, applicability to driver safety, and feature purchase interest received relatively high positive responses.

Message Interpretation with Intersection Collision Danger Scenario

Table 3 outlines the response categories derived when participants were asked to provide an open-ended interpretation of the DVI (i.e., visual alert icon) after experiencing this feature demonstration scenario. Consistent with the message interpretability ratings discussed above for this alert (i.e., "Icon communicates message?"), participants interpreted the message as intended. This suggests the driver would pay attention to the crossing vehicle and exhibit caution.

Table 3. Intersection Collision Danger Message Categorization

Message Categories	Percent Response
Car coming through intersection/intersection hazard	35%
Car rapidly approaching/Car approaching from right	27%
Possible collision/Collision course/Crash likely	18%
Other: Car failing to stop/cross traffic/car pulling out, etc	18%

HARD BRAKING AHEAD

As shown in Table 4, 60% of the participants (ranging between 52% - 60% across groups) rated the timing of the hard braking ahead alert as occurring at the right time, with 40% suggesting the alert timing was "too late". This latter bias, similar to the Time of Day effect observed in the intersection incursion scenario, may be related once again to the perceived urgency of this scenario and the participant's desire for additional time to respond to this scenario. The reader should be reminded that the alert in this scenario was issued as soon as the (lead) scenario vehicle exceeded 0.3 g's. Therefore, from an alert timing perspective, there was no opportunity to issue the alert any sooner (given this 0.3 g braking criterion) than what drivers experienced in the current study. Furthermore, it should be noted that lowering the threshold for triggering this alert (which would have made the alert come on sooner) could be problematic from a nuisance alert (or false alarm) perspective. The icon

Table 4. Percentage breakdown of responses to the post scenario questions for the hard braking event.

Question	Gender		Daytime/Nighttime	
	Male	Female	Day	Night
In my opinion, the message appeared: - At the right time (Too Late)	58% (39%)	58% (41%)	60% (37%)	52% (45%)
The icon clearly communicates the intended message? - Agree (Disagree)	40% (35%)	48% (34%)	46% (33%)	40% (39%)
The feature provides useful information? - Agree (Disagree)	73% (13%)	61% (13%)	68% (11%)	64% (17%)
The feature would increase my driving safety? - Agree (Disagree)	76% (12%)	64% (11%)	83% (7%)	64% (19%)
I want this feature on my next vehicle? - Agree (Disagree)	62% (16%)	59% (13%)	62% (11%)	57% (22%)

(Significant difference between groups p < 0.05)*

interpretation received an overall low rating across all the groups, however, participants deemed this feature to be useful and enhance driver safety. The remaining questions shown in Table 4 addressing icon usefulness, applicability to driver safety, and feature purchase interest received somewhat to relatively high positive responses.

Message Interpretation with Hard Braking Scenario

Table 5 shows the categorization of the open-ended responses for this feature demonstration scenario. Consistent with the message interpretability ratings discussed above for this alert, participants did not generally interpret the message exactly as intended. That said, the majority of participants provided a response suggesting the driver would pay attention to the forward traffic scene due to a braking vehicle or hazard ahead and exhibit caution. These results, when coupled with the message interpretability ratings, suggest that efforts to further develop this "hard breaking ahead" or perhaps use a related icon for this feature (e.g., a Forward Collision Warning icon) could be beneficial.

Table 5. Hard Braking Ahead Message Categorization

Message Categories	Percent Response
Hard braking ahead/braking hard/car stopping	31%
Turning vehicle/right turning vehicle/turning off road	28%
Other: Same icon/not sure/message too quick/no comment	22%
Vehicle hazard/caution ahead/hazard ahead	17%
Vehicle behaving erratically	2%

ELECTRONIC STABILITY CONTROL ACTIVATED AHEAD

As shown in Table 6, 57% of the participants (ranging between 51% - 59% across groups) rated the timing of the ESC activated ahead alert as occurring at the right time. The rating, similar to that observed in the previously discussed intersection incursion and hard braking ahead scenarios, may be related to the perceived urgency of this scenario and the participant's desire for additional time to respond to the scenario. Similar to what was discussed with the hard braking ahead alert, the reader should be reminded that in practice this alert could only be issued after an activation of the ESC system in a lead vehicle. Therefore, from an alert timing perspective, there may be little (if any) opportunity to issue the alert any sooner than what drivers experienced in the current study. Furthermore, it should be noted that lowering the threshold for triggering this alert (which would have made the alert come on sooner) could be problematic from a nuisance alert (false alarm) perspective. The remaining questions shown in Table 6 addressing icon usefulness, applicability to driver safety, and feature purchase interest received somewhat to relatively high positive responses.

A significant Time of Day effect was found under this scenario ($\chi 2(1, N=99) = 5.36, p < 0.05$), under which the nighttime group disagreed with this alert timing statement more often than the daytime group. This finding may be due to the added difficulty nighttime participants had observing this scenario. If participants failed to detect the tire and therefore the reason for the ESC activation, their interpretation of the scenario may have been affected.

Message Interpretation with Electronic Stability Control Scenario

As Table 7 shows the categorization of the open-ended responses for this feature demonstration scenario, and indicates that participants did not interpret the message exactly as intended. However, the vast majority of participants provided a response suggesting that the driver would pay attention to the forward traffic scene due to a vehicle ahead losing control or hazard ahead and exhibit caution.

Table 6. Percentage breakdown of responses to the post scenario questions for the ESC event.

Question	Gender		Daytime/Nighttime	
	Male	Female	Day	Night
In my opinion, the message appeared: - At the right time (Too Late)	56% (41%)	57% (40%)	59% (38%)	51% (46%)
The icon clearly communicates the intended message? - Agree (Disagree)	55% (23%)	58% (26%)	60%* (17%)	49%* (39%)
The feature provides useful information? - Agree (Disagree)	75% (6%)	74% (10%)	73% (2%)	76% (19%)
The feature would increase my driving safety? Agree (Disagree)	73% (8%)	69% (15%)	73% (9%)	66% (17%)
I want this feature on my next vehicle? - Agree (Disagree)	69% (11%)	64% (13%)	65% (11%)	68% (15%)

(Significant difference between groups p < 0.05)*

Table 7. Electronic Stability Control Message Categorization

Message Categories	Percent Response
Vehicle swerving/skidding ahead, slippery/stability problems	45%
Object/Obstacle/Tire/Road Hazard/Something on the road	42%
Motion ahead/curved road ahead/wasn't sure/didn't see	12%

Table 8. Percentage breakdown of responses to the post scenario questions for the Post-Crash Notification event.

Question	Gender		Daytime/Nighttime	
	Male	Female	Day	Night
In my opinion, the message appeared: - At the right time (Too Late)	90% (10%)	89% (5%)	91% (7%)	86% (7%)
The icon clearly communicates the intended message? - Agree (Disagree)	64% (24%)	76% (16%)	79%* (11%)	51%* (37%)
The feature provides useful information? - Agree (Disagree)	86% (5%)	88% (2%)	92% (2%)	80% (5%)
The feature would increase my driving safety? - Agree (Disagree)	87% (7%)	93% (3%)	92% (4%)	75% (7%)
I want this feature on my next vehicle? - Agree (Disagree)	80% (8%)	82% (3%)	84% (5%)	85% (7%)

(Significant difference between groups p < 0.05)*

POST-CRASH NOTIFICATION

As shown in Table 8, 90% of the participants (ranging between 86% - 91% across groups) rated the timing of the post-crash notification alert as occurring at the right time. 75% of participants (ranging between 51% - 79% across groups) agreed with the statement that the post-crash notification alert communicated the intended message. In addition, a significant Time of Day effect was found ($\chi 2(1, N=112) = 12.98$, $p < 0.01$), under which the daytime group agreed with this message interpretation statement more often than the nighttime group (79% versus 51%). Once again, this finding may be due to the added difficulty nighttime participants had observing this scenario. Furthermore, if participants failed to detect the construction barrel in this scenario, they may have misinterpreted the scenario vehicle as someone that had simply pulled off the roadway. The remaining questions shown in Table 8 addressing icon usefulness, applicability to driver safety, and feature purchase interest received relatively high positive responses.

Message Interpretation with Post Crash Notification Scenario

Table 9 shows the categorization of the open-ended responses for this feature demonstration scenario. Again, there was a range of message interpretations associated with this scenario. However, similar to previous icon presentations,

Table 9. Advanced Collision Notification Message Categorization

Message Categories	Percent Response
Obstacle/Blockage/Hazard/Something in the road	50%
Accident/Collision/Crash Ahead	29%
Car broken down ahead/Car disabled/vehicle stopped	14%
Danger ahead/vehicle spun out/not sure	6%

Table 10. Percentage breakdown of responses to the post scenario questions for the slow moving vehicle event.

	Gender		Daytime/Nighttime	
Question	Male	Female	Day	Night
In my opinion, the message appeared: - At the right time (Too Late)	73%* (16%)	86%* (5%)	77% (10%)	84% (12%)
The icon clearly communicates the intended message? - Agree (Disagree)	56% (16%)	58% (26%)	79% (21%)	61% (21%)
The feature provides useful information? - Agree (Disagree)	84% (7%)	70% (5%)	76% (6%)	77% (5%)
The feature would increase my driving safety? - Agree (Disagree)	84% (6%)	76% (9%)	82% (11%)	77% (2%)
I want this feature on my next vehicle? - Agree (Disagree)	69% (11%)	69% (10%)	68% (11%)	70% (10%)

(Significant difference between groups $p < 0.05$)*

participants suggested they would orient their attention towards the warning and perhaps would increase vigilance towards the outside driving environment in an effort to anticipate a potential hazard.

SLOW MOVING VEHICLE AHEAD

As shown in Table 10, 80% of the participants (ranging between 73% - 86% across groups) rated the timing of the slow moving vehicle ahead alert as occurring "at the right time". In addition, a significant Gender effect was observed, ($\chi 2(1, N=112) = 4.26, p < 0.05$), with more males rating the alert as occurring "too late" compared to females (16% versus 5%). The remaining questions shown in Table 10 addressing icon usefulness, applicability to driver safety, and feature purchase interest received somewhat to relatively high positive responses.

Message Interpretation with Slow Moving Vehicle Ahead Scenario

Table 11 shows the open-ended interpretation responses given by participants for this scenario. This scenario (which employed an icon also used in other scenarios), was interpreted correctly in most cases, with over half of the participants interpreting the exact scenario meaning, and another 19% comprehending that it was warning about an event ahead.

Table 11. Slow moving vehicle message categorization

Message Categories	Percent Response
Slow moving vehicle/slow vehicle/stopping vehicle	33%
Vehicle pulling off/Vehicle on the shoulder	26%
Caution/Hazard ahead	19%
Something on road in front/vehicle broken down	16%
Not sure/Heard the auditory warning only	4%

Vehicle Ahead with Hazard Warning Flashers Activated

As shown in Table 12, 92% of the participants (ranging between 86% - 97% across groups) rated the timing of the hazard warning flashers ahead alert as occurring "at the right time". A Gender effort indicated that males were more likely to rate the alert as occurring "too late" compared to females, $\chi 2(1, N=120) = 6.02, p <0.05$); 10% versus 2%, respectively. In addition, a Time of Day effect indicated that nighttime participants were more likely to rate the alert as occurring "too late" (12%) compared to daytime participants (2%), ($\chi 2(1, N=121) = 4.67, p < 0.05$). The remaining questions shown in Table 12 addressing icon usefulness, applicability to driver safety, and feature purchase interest received somewhat to relatively high positive responses.

Table 12. Percentage breakdown of responses to the post scenario questions for the hazard warning flashers ahead event.

Question	Gender		Daytime/Nighttime	
	Male	Female	Day	Night
In my opinion, the message appeared: - At the right time (Too Late)	87%* (10%)	97%* (2%)	95%* (2%)	86%* (12%)
The icon clearly communicates the intended message? - Agree (Disagree)	67% (17%)	63% (19%)	71%* (13%)	54%* (28%)
The feature provides useful information? - Agree (Disagree)	81% (9%)	75% (0%)	81% (2%)	73% (9%)
The feature would increase my driving safety? - Agree (Disagree)	80% (9%)	77% (3%)	80% (6%)	75% (7%)
I want this feature on my next vehicle? - Agree (Disagree)	65% (15%)	65% (5%)	62% (9%)	70% (11%)

(Significant difference between groups $p < 0.05$)*

As shown in Table 12, 79% of the participants (ranging between 54% - 71% across groups) agreed with the statement that the hazard warning light activated ahead alert communicated the intended message. This finding could be related to the fact the alert icon used for this alert was also used for both the hard braking ahead and slow moving vehicle ahead scenarios. A significant Time of Day effect was observed, ($\chi2(1, N=104) = 4.54$, $p < 0.05$), with the nighttime group having a higher number of participants disagreeing with this statement compared to the daytime group (28% versus 13%). This finding may be due to the added difficulty participants may have had observing this scenario under daytime conditions (e.g., detecting the hazard lights being activated). The remaining questions shown in Table 12 addressing icon usefulness, applicability to driver safety, and feature purchase interest received somewhat to relatively high positive responses.

Message Interpretation with Vehicle Ahead with Hazard Warning Flashers Activated Scenario

Table 11 shows the open-ended interpretation responses given by participants for this scenario. Results indicate that the icon (which employed an icon also used in other scenarios) was well-interpreted as indicating caution ahead involving another vehicle (e.g., vehicle hazard).

Table 13. Vehicle ahead with hazard warning flashers activated message categorization

Message Categories	Percent Response
Car broken down/disabled/stopped on the side of road	46%
Other: Car on right/flat tire/changing tire/not sure	16%
Caution/Hazard ahead	15%
Vehicle Hazards/Vehicle with hazards on	14%

Driver and Passenger Alert Timing Responses

Alert timing ratings (e.g., "Just right" or "Too late") were also analyzed based on whether participants were passengers or drivers across the scenarios. The third category, a "Too early" response, was removed from the analysis based on a low response rate across the majority of the events. Therefore, group comparisons were only made between drivers and passengers based on if they rated the alert as "Just right" or "Too late". As can be seen in Figure 4, results indicated that there were no significant differences between the driver and passenger groups ($p > 0.05$) in their ratings of the alert timings. Although the magnitude of the differences between some of the events (e.g., Hard braking, ESC, and PCN) suggest potential differences between participant role based on their location, the chi-square tests failed to reach marginal significance ($p > 0.60$). From a testing methodology perspective, these results suggest that, participant role/ location in the vehicle (driver or passenger) did not impact respondent's judgments with respect to alert timeliness. This finding may be useful to exploit in further research with respect to gathering similar data more efficiently.

POST DRIVE QUESTIONNAIRE

The post-drive questionnaire was administered after the participants had experienced all the V2V feature demonstrations. This questionnaire examined participant interest in V2V, their attitudes towards the technology, their preferences for the DVI, and some marketing questions not covered in these results. The results are generally presented using the entire sample, although comparisons were made between the Gender and Time of Day groups where appropriate.

General Interest in V2V System

The first item of the post-drive questionnaire asked participants to indicate how interested they would be in having V2V technology on their vehicle. As shown in Figure 5, 84% of the participants responded positively to this question. In addition, note that only one of the 125 participants reporting they were not interested in the system.

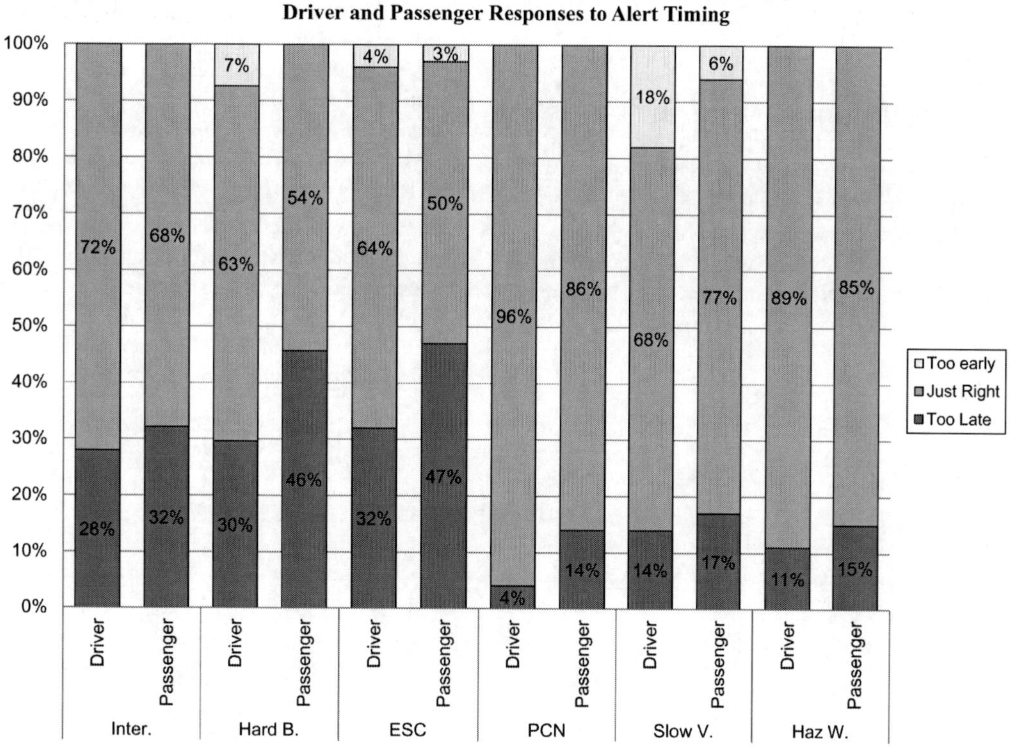

Figure 4. Alert response judgments by Drivers and Passengers

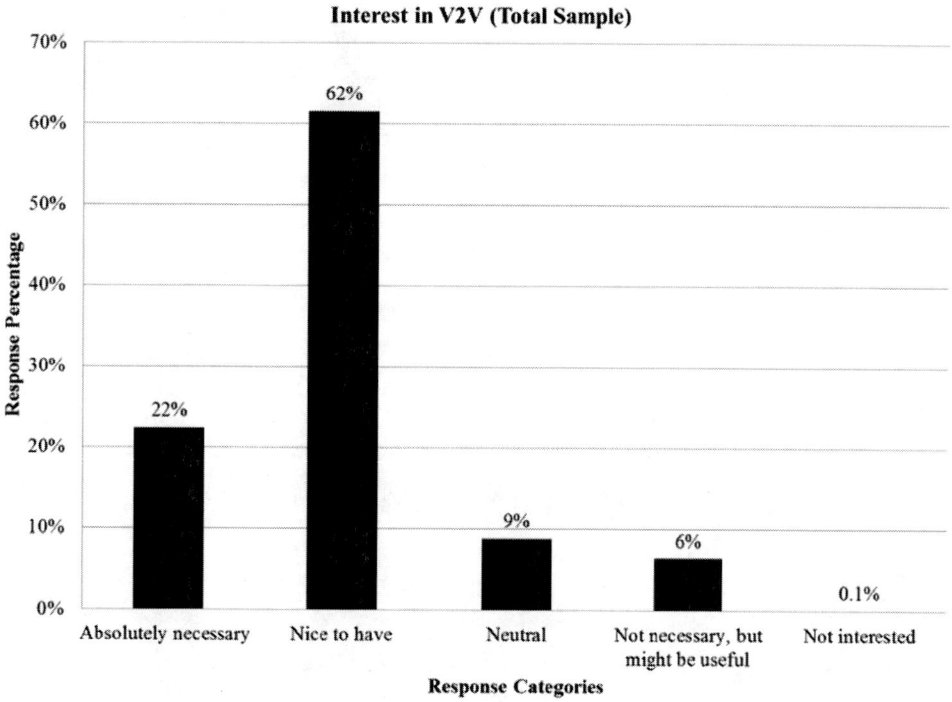

Figure 5. Overall Interest in the V2V System

Technology Deployment

The second item of the post-drive questionnaire asked participants to select a percent/ratio of vehicles (e.g., "<10% or less than 10 out of 100 vehicles on the road") on the roadway at which point they would feel the technology would be worth having on their vehicle, while keeping in mind that V2V can only communicate with similarly equipped V2V vehicles. It should be stressed that drivers were not give any system price assumptions in this question. As shown in Figure 6, 42% of participants felt that a deployment level of at least 50% was required before participants felt the technology was worth having, and 76% of the participants felt that a deployment level of at least 75% was required before they felt the technology was worth having. Only one participant responded that technology would not be worth having irrespective of deployment level.

Attitudes towards V2V

In the next series of the post-drive questions, participants were asked to rate various statements on a 5-point Likert scale (identical to the 1-Strongly Disagree through 5= Strongly Agree, as discussed earlier) addressing their general attitudes towards the V2V system and the V2V features they experienced (including the DVI) during the test drivers.

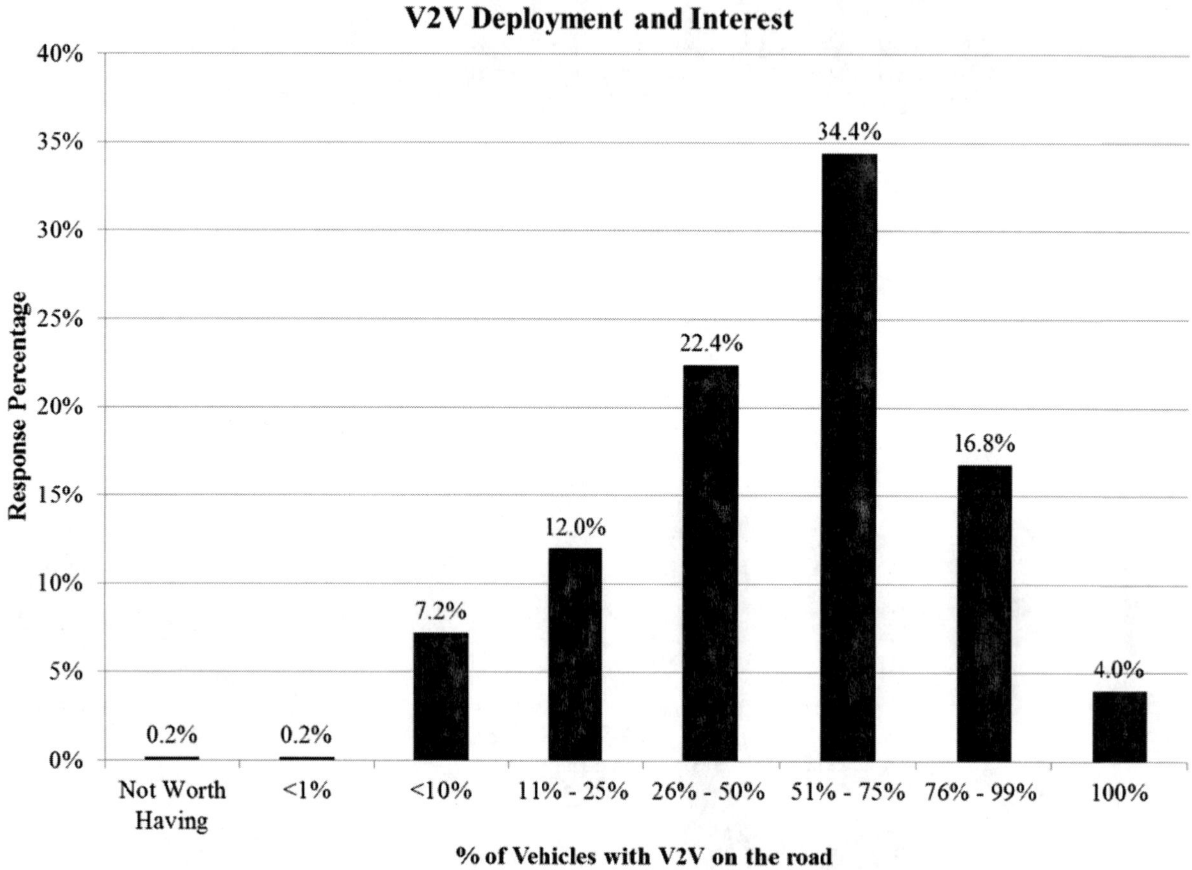

Figure 6. Deployment levels and V2V technology interest

Message Information and Safety

One set of questions asked participants to rate their overall impression of whether the messages were clearly communicated, if the system would be helpful for their driving, if the system would increase their driving safety, and if they would feel safer driving with V2V technology. As shown in Figure 7, a larger majority of participants agreed that the V2V system would be helpful (78%) and increase their driving safety (82%), and the majority of participants felt the system would make them feel safer (59%) and messages were clearly communicated (58%). 16% disagreed with this latter assertion. Consistent with the Time of Day effects observed in the in-vehicle scenarios response questionnaires, a significant Time of Day effect was also observed for the message clarity statement ($\chi2(1, N=89) = 5.81$, $p < 0.05$), under which daytime participants agreed more often than nighttime participants that the messages were clearly conveyed (85% versus 65%). As was discussed earlier, the lower message clarity ratings observed during nighttime relative to daytime driving conditions is likely to be related to the added difficulty participants may have had observing the demonstration scenario under nighttime conditions.

Driver Vehicle Interface

Another set of items in the post-drive questionnaire asked participants to rate statements addressing more directly the DVI they experienced. These results are shown in Figure 8. When asked if the visual display (i.e., icon) was sufficient for alerting the driver without the addition of the beeping alert, 74% of the participants disagreed with this statement. Similarly, when asked if both the visual icon and beeping alert were sufficient for alerting the driver (which was the alerting approach used in this study), 79% of the participants agreed with this statement. Finally, when asked if this feature would be annoying, 72% of the participants disagreed with this statement, with 22% remaining neutral toward this statement. It should be noted that these annoyance results should be treated cautiously since it is difficult to estimate how often V2V alerts may occur during real world driving. Overall, these results support the use of a multi-modality V2V DVI approach for the features evaluated in the current study.

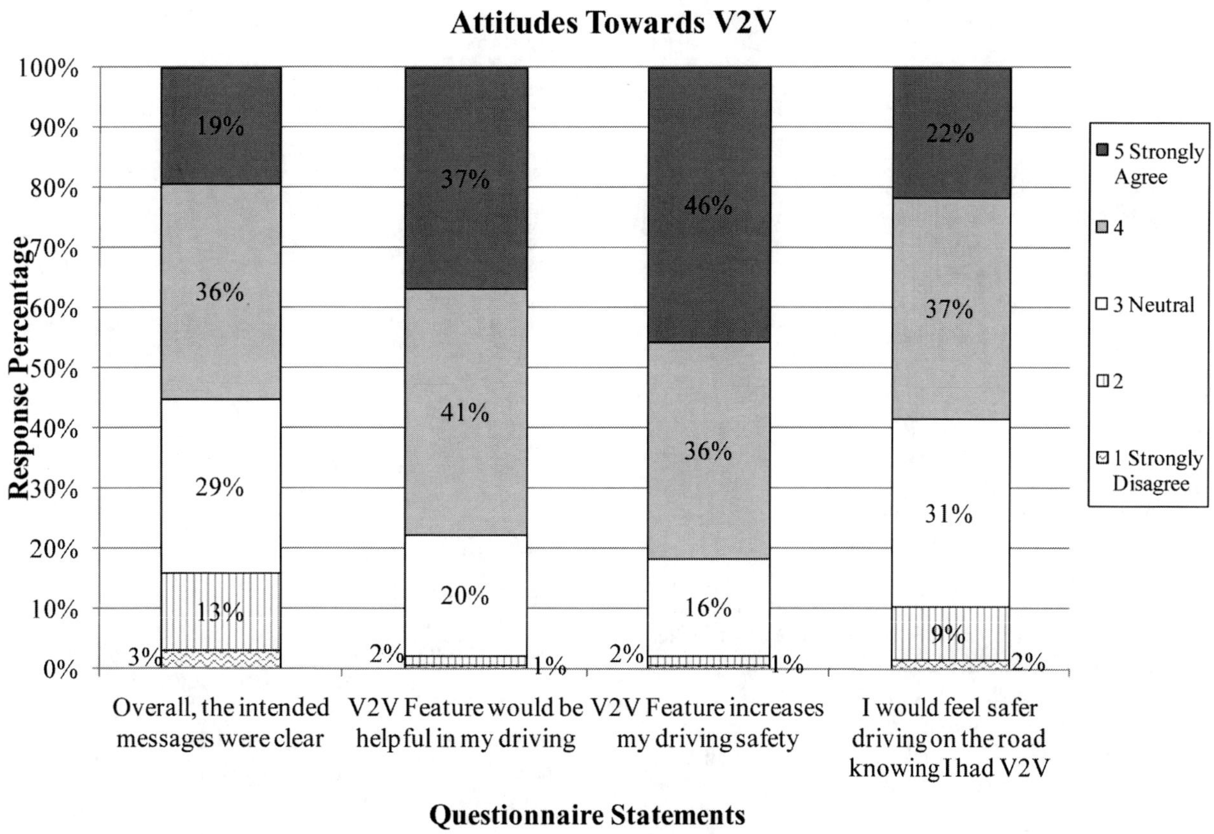

Figure 7. Message Clarity and Driving Safety with V2V

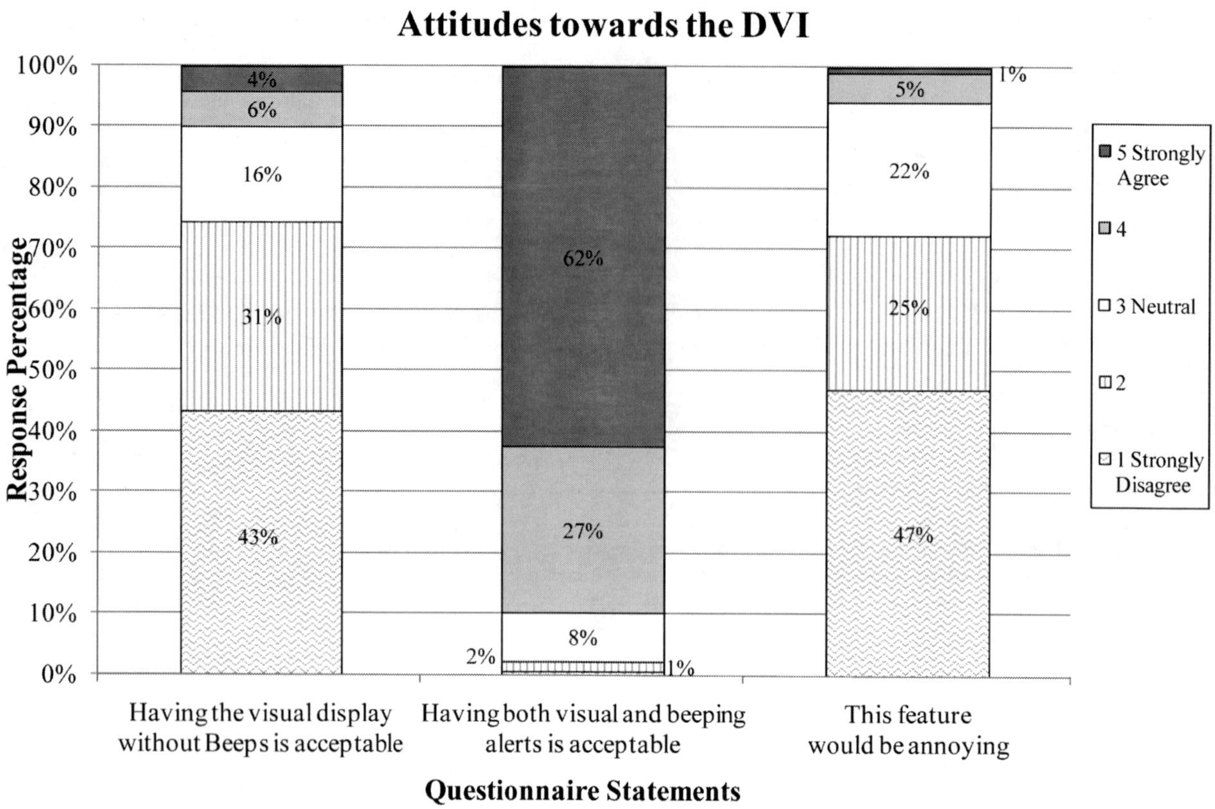

Figure 8. General attitudes towards the DVI and features

V2V Driver Vehicle Interface Location

Another set of items in the post-drive questionnaire asked participants to rate the acceptability of various V2V visual alert display locations. Figure 9 provides the diagram participants were provided for this display rating. As noted above, during the study the visual alerts (icons) were displayed to the driver (and passengers) at a "top of dash" location (as shown previously in Figure 1), which could potentially bias the pattern of driver preferences. Furthermore, it should be stressed that that this display approach was not production representative, most notably in terms of the large display size and the add-on appearance (i.e. lack of "integration into the dash" aesthetics).

As shown in Figure 10, participants were decidedly less favorable toward placing V2V display information on either the center area or speedometer locations, with 16% and 41% of the participants, respectively, agreeing this was an acceptable location. In addition, for these center area and speedometer locations, 57%% and 33% of the participants, respectively, disagreed this was an acceptable location. In contrast, participants responded favorably toward placing V2V information on either a head-up display (HUD) or "top of dashboard" display location, with the majority of participants (65% and 56%, respectively) agreeing this was an acceptable location, and roughly 20%-25% of the participants considering these two display locations as unacceptable. This latter finding may be due to the drivers lacking familiarity with HUD technology and the particular (non-production representative) "top of dashboard" display study participants experienced.

Overall, these subjective results suggest that V2V information should be placed "high" on either a HUD or "top of dashboard" location rather than a center area or speedometer (i.e., instrument panel) location. Such a recommendation is consistent with results from recent driver performance studies that have examined the driver performance implications of a Forward Collision Warning (FCW) visual alert location [1,2]. These recent driver performance studies also support the SAE J2400 [3] recommendation advising against the use of the instrument panel location for the FCW visual alert.

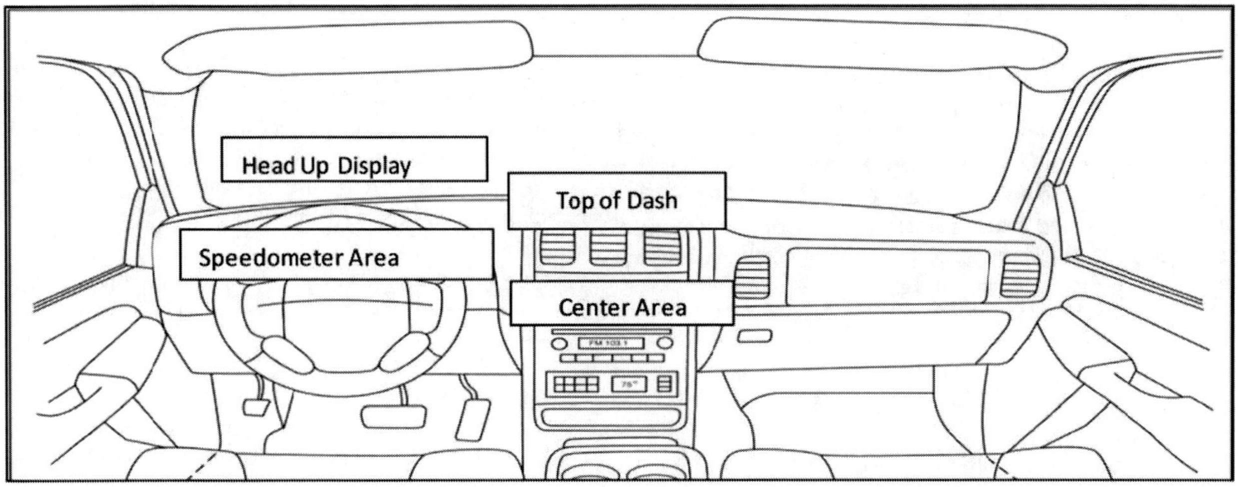

Figure 9. Driver Vehicle Interface Location Descriptions

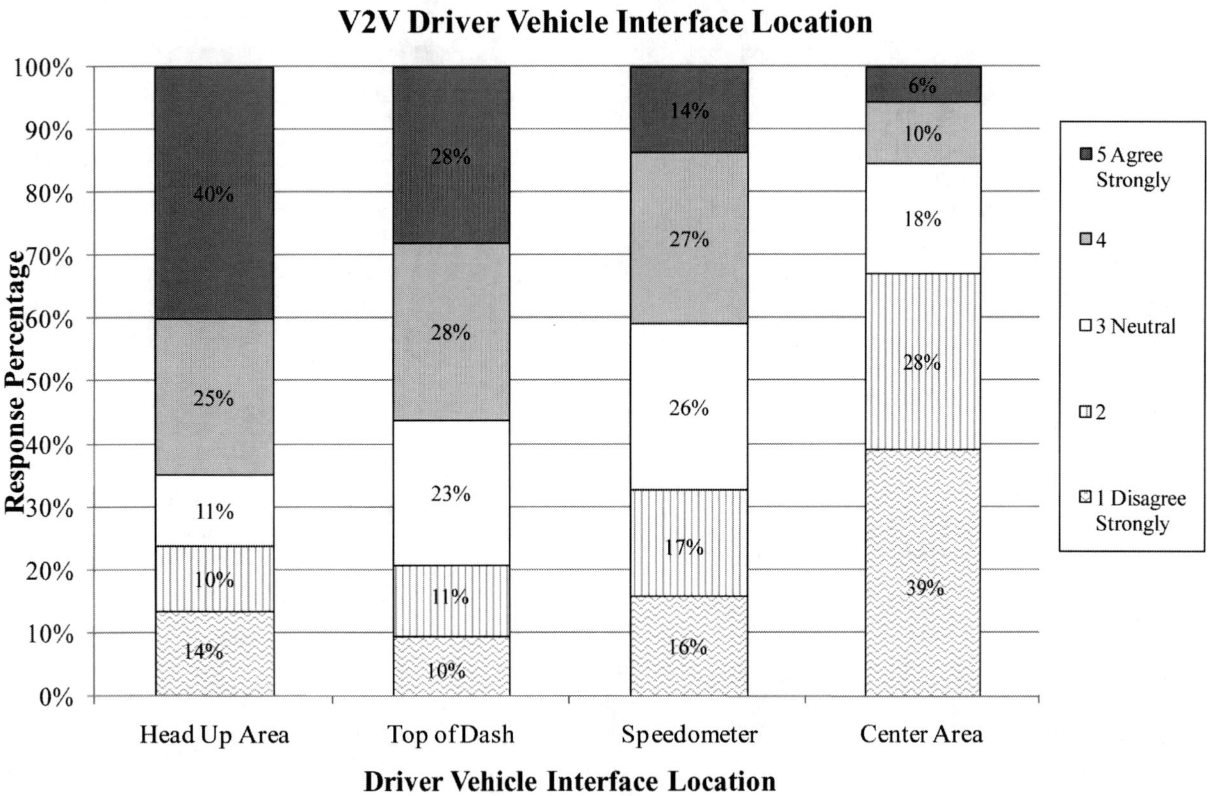

Figure 10. V2V Driver Vehicle Interface Specifications

Technology Adoption

Another set of items in the post-drive questionnaire asked participants to rate statements relating to having their future or current vehicle equipped with V2V technology, and the importance of having other V2V-equipped vehicles identified. The results for these questions are shown in Figure 8. When asked if they would be one of the first to own V2V technology, 20% of the participants agreed with this statement, with 38% of participants remaining neutral. When asked if there will ever be enough vehicles equipped with V2V technology to make the technology useful, 45% of the participants agreed with this statement, with 37% of participants remaining neutral. When asked if they would have V2V technology installed on their current vehicle, 24% of the participants agreed with this statement, with 26% of participants remaining neutral. However, when asked if they would purchase V2V technology on their next vehicle, 48% of the participants agreed with this statement, with 30% of participants remaining neutral. Finally, when participants were asked if it was important that other V2V-equipped vehicles could be identified (or "badged") around them while they were driving (e.g., with a colored antennae), the nature of the responses suggested participants were generally neutral toward this idea (with an overall slight bias toward disagreeing with this statement).

Overall, results from this series of questions suggest that drivers would prefer to have V2V technology on their next vehicle (rather than having it retrofitted on their current vehicle). In addition, the relatively high level of "neutral" ratings across this series of questions may be due to the uncertainty participants may have surrounding the future level of V2V technology deployment, as well as the difficulty of predicting how often V2V alerts may occur during real world driving (and hence, the associated usefulness of these V2V features).

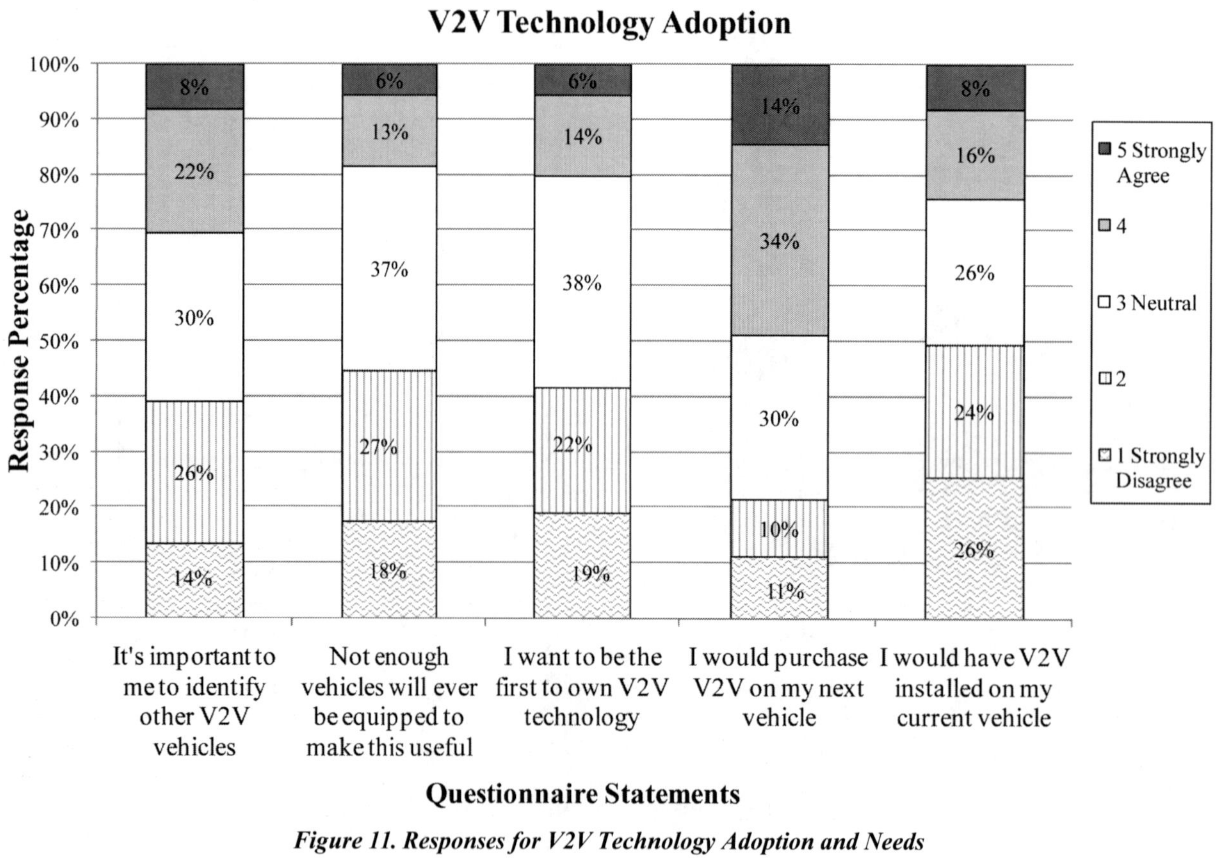

Figure 11. Responses for V2V Technology Adoption and Needs

SUMMARY

The purpose of this test track study was to gather data on driver perceptions and opinions with respect to features associated with a prototype Vehicle to Vehicle (V2V) communication system, the driver vehicle interface (DVI) approach associated with these features (e.g., the visual alert icons) and V2V technology as a whole. This study employed a unique "feature demonstration" approach using platoons of V2V-equipped vehicles, and gathered data from 125 test participants (males and females) under both daytime and nighttime driving conditions. The visual alert icons associated with the various V2V features demonstrated were based on previous General Motors icon comprehension/rating research, and were located on a "top of the dashboard" add-on display positioned at vehicle centerline. These icons were accompanied by auditory beeps that are commonly employed as automotive collision avoidance-related auditory warnings.

Test participants were provided a brief overview of V2V technology prior to six feature demonstration drives, but were not provided advanced information on the specific nature of the V2V features prior to experiencing them during these drives (as either a driver or passenger). This latter experimental strategy was employed to enable determining the extent to which the alert approach evaluated provided an intuitive explanation of the scenario, and to loosely approximate what drivers might experience when initially experiencing V2V technology features. The features demonstrated included alerting the test participants to the following events ahead of the vehicle: vehicle braking hard, slow moving vehicle, vehicle with hazard warning lights activated, post-crash notification, electronic stability control (ESC) activation, and a potential intersection crash situation.

Overall, the majority of participants indicated that the V2V features and technology they experienced provided important safety-related information about the status of the roadway ahead. The V2V visual alert icons associated with the features demonstrated were generally perceived as meaningful and interpreted correctly. In cases where the icon associated with the feature was not interpreted exactly as intended, the open-ended icon comprehension responses suggested the driver, at a minimum, would pay attention to the forward traffic scene and exhibit caution to events unfolding ahead of them. Furthermore, with continued experience, it can be reasonably expected that icon comprehension would increase beyond that observed during this initial, one time experience with each of these features. These results also provide support for the merit of the approach used in the current study of using the same icon for multiple V2V features (e.g., vehicle braking hard, slow moving vehicle, hazard warning lights activated), which may be advantageous given potential display packaging challenges and the multitude of V2V features currently under deployment consideration. The results also supported the use of a multi-modality V2V DVI approach (including both visual auditory alert components) for the features evaluated in the current study, as well as employing a head-up display (HUD) or "high" head-down display location for the V2V visual alerts (rather than an instrument panel or center stack location). This latter subjective data finding is consisted with recent driver performance studies addressing the Forward Collision Warning (FCW) visual alert location [1,2] and SAE J2400 [3].

The timing of the V2V alert associated with the features demonstrated was generally rated either as "just right" or "too late", and no difference was found on these ratings between driver and passenger participants. In addition, there was an increased tendency for "too late" ratings when the perceived urgency of the scenario dictated participants' desire for additional time to respond (i.e., potential intersection collision, vehicle braking hard, or ESC activation ahead). In some cases these "too late" ratings were further exaggerated under nighttime conditions, where participants may have had additional difficulty observing the scenario unfold. It should be noted that there may be little (if any) opportunity for a V2V system to issue the alert any sooner than what drivers experienced in these scenarios, and that lowering the threshold for triggering the alert in the V2V scenarios demonstrated (which would have made the alert come on sooner) could be problematic from a nuisance alert (false alarm) perspective.

Across the six V2V features demonstrated, the large majority of the participants (ranging from 70% to 86% across features) indicated that the feature would increase driving safety and that the feature was useful (ranging from 66% to 88% across features). It should be noted that the purchase interest consideration results observed here should be treated with appropriate levels of caution due to challenges participants faced when making inherent assumptions about the future level of V2V technology deployment and how often the V2V alerts they experienced may occur during real world driving. Sixty-eight percent of the participants agreed that V2V deployment levels across the fleet needed to between 51% and 75% before they would consider purchasing the technology, whereas 90% of the participants agreed that deployment levels needed to be between 76% and 99% before they would consider purchasing the technology. Finally, participants preferred to have V2V on their next vehicle rather than having it retrofitted on their current vehicle, and did not believe it was important that other V2V-equipped vehicles could somehow be identified around them while they were driving.

In conclusion, the subjective data gathered under this study effort suggest that the V2V features examined, as well as the V2V system as a whole, were well-received and perceived as providing safety benefits. These results, coupled with the relative affordability of V2V compared to other autonomous sensing systems, suggest that the V2V system may with anticipated future levels of deployment provide a promising approach to improve traffic safety and increase the penetration of safety systems across the vehicle fleet. In addition, these results suggest that the DVI and customer education approach surrounding the introduction of V2V technology are likely to play a key role in enabling driver acceptance and ensuring that drivers take full advantage of the wide range of potential driver safety and convenience benefits offered by this emerging technology.

REFERENCES

1. Lind, H., "An Efficient Visual Forward Collision Warning Display for Vehicles," SAE Technical Paper 2007-01-1105, 2007, doi:10.4271/2007-01-1105.

2. Perez, M., Kiefer, R., Haskins, A., and Hankey, J., "Evaluation of Forward Collision Warning System Visual Alert Candidates and SAE J2400," *SAE Int. J. Passeng. Cars-Mech. Syst.* **2**(1):750-764, 2009, doi:10.4271/2009-01-0547.

3. SAE International Surface Vehicle Information Report, "Human Factors in Forward Collision Warning Systems: Operating Characteristics and User Interface Requirements," SAE Standard J2400, Issued Aug. 2003.

CONTACT INFORMATION

Christopher J. Edwards
cedwards@vtti.vt.edu

or

Dr. Raymond Kiefer
raymond.j.kiefer@gm.com

ACKNOWLEDGMENTS

General Motors provided the funding and the V2V test fleet for this study.

SAE International®

Vehicle Safety Communications - Applications: System Design & Objective Testing Results

2011-01-0575
Published
04/12/2011

Farid Ahmed-Zaid
Ford Motor Company

Hariharan Krishnan
General Motors Company

Michael Maile
Mercedes Benz REDNA

Lorenzo Caminiti
Toyota Motor Engineering & Mfg NA Inc.

Sue Bai
Honda R&D Americas Inc.

Steve VanSickle
Danlaw, Inc.

doi:10.4271/2011-01-0575

ABSTRACT

The USDOT and the Crash Avoidance Metrics Partnership-Vehicle Safety Communications 2 (CAMP-VSC2) Consortium (Ford, GM, Honda, Mercedes, and Toyota) initiated, in December 2006, a three-year collaborative effort in the area of wireless-based safety applications under the Vehicle Safety Communications-Applications (VSC-A) Project. The VSC-A Project developed and tested communications-based vehicle safety systems to determine if Dedicated Short Range Communications (DSRC) at 5.9 GHz, in combination with vehicle positioning, would improve upon autonomous vehicle-based safety systems and/or enable new communications-based safety applications. The project addressed the following objectives:

- Assess how previously identified crash-imminent safety scenarios in autonomous systems could be addressed and improved by DSRC+Positioning systems
- Define a set of DSRC+Positioning based vehicle safety applications and application specifications including minimum system performance requirements
- Develop scalable, common vehicle safety communication architecture, protocols, and messaging framework (interfaces) necessary to achieve interoperability and cohesiveness among different vehicle manufacturers. Standardize this messaging framework and the communication protocols (including message sets) to facilitate future deployment.
- Develop requirements for accurate and affordable vehicle positioning technology needed, in conjunction with the 5.9 GHz DSRC, to support most of the safety applications with high-potential benefits
- Develop and verify a set of objective test procedures for the vehicle safety communications applications

In this paper, we summarize the work that took place in the VSC-A Project in the areas of system design and objective testing. We first introduce the VSC-A system framework. We then list the crash imminent scenarios addressed by the VSC-A Project and the safety applications selected to potentially address them. Next we describe the VSC-A test bed system development. This test bed was ultimately used to verify Vehicle-to-Vehicle (V2V) communication interoperability between Ford, GM, Honda, Mercedes-Benz, and Toyota vehicles. Public demonstrations of V2V interoperability were held in New York City at the 2008 Intelligent Transport Systems (ITS) World Congress. The test bed also served to validate the system and minimum performance specifications that were developed as part of this project. We discuss one of the most important achievements of the project in the communication area, i.e., implementation, testing, verification, and standardization of a safety message that supports all of the VSC-A safety applications. The result is the Basic Safety Message (BSM) as defined in the SAE J2735 Message Set Dictionary standard. Details of the objective test procedures are presented next and are followed by a summary of the performed test scenarios (test descriptions, speeds, number of runs for each test, type of test, etc.) with the corresponding objective testing results. We conclude the paper with a section summarizing the accomplishments of the project and also identify potential next steps and recommendations based on the technical results and engineering experience gained throughout the execution of the VSC-A Project.

INTRODUCTION

Vehicle-to-Vehicle (V2V) safety communications can play a major role in addressing vehicle crashes where multiple vehicles are involved. According to [1], this technology can reduce, mitigate, or prevent 82 percent of crashes by unimpaired drivers. The communications technology for V2V is 5.9 GHz Dedicated Short Range Communications (DSRC). This wireless communications technology has a very low latency and is considered to be the technology of choice for the types of crash avoidance applications that were prototyped in the Vehicle Safety Communications-Applications (VSC-A) Project [2]. The major objectives of the VSC-A development activities were the:

- Selection of high-value safety applications
- Development of a test bed that allowed interoperability between different car manufacturers
- Development and standardization of a message set for vehicle safety communications
- Development of an accurate relative positioning system
- Prototyping of safety applications
- Objective testing of the safety applications

A primary goal of the VSC-A Project was to determine whether systems that utilized DSRC-based V2V communications and positioning can help overcome limitations of autonomous systems and enhance the overall performance of safety systems. One potential advantage of V2V safety communications is that it may provide significant, additional information about the driving situation and expand the awareness horizon of the vehicle well beyond the capabilities of vehicle-autonomous sensors. Another advantage of V2V systems is that it may be possible to integrate such systems on vehicles in which the system was not original equipment, including retrofit of existing vehicles.

In order to gauge the feasibility of such systems, a reference system and applications to address crash imminent scenarios were implemented. This reference system ("test bed") combined communications, accurate relative positioning and security and was integrated with the vehicles from the five Original Equipment Manufacturers (OEMs) that participated in the VSC-A Project. A fundamental aspect of the project was the establishment of interoperability between different OEMs. This interoperability requirement led to the development of the V2V message set, which was standardized in SAE J2735 as the Basic Safety Message (BSM) [3]. The development of the test bed and the applications followed a systems engineering process and the resulting minimum performance requirements formed the basis for the development and the testing of the applications. To test the performance of the test bed and the applications, objective test procedures were developed together with the United States Department of Transportation (USDOT) and the testing was performed at the Transportation Research Center (TRC) in East Liberty, Ohio with the aid of the National Highway Traffic Safety Administration's (NHTSA) Vehicle Research and Test Center (VRTC).

CRASH SCENARIOS AND APPLICATION SELECTION

To provide a foundation for the VSC-A Project, the USDOT evaluated pre-crash scenarios based on the 2004 General Estimated Systems (GES) crash database. This list served as the basis for the selection of the safety applications to be prototyped under the VSC-A Project. Each crash scenario was assigned a composite crash ranking determined by taking the average of the crash rankings by frequency, cost, and functional years lost for each scenario. The crash scenarios were then sorted based on the composite ranking and were analyzed to evaluate whether autonomous safety systems and/ or vehicle safety communications would offer the best opportunity to adequately address the scenarios.

From this ranked list of crash scenarios (based on crash frequency, crash cost and functional years lost) the top seven (7) crash scenarios to be addressed by the VSC-A Project were selected. The selected crash-imminent scenarios were analyzed and potential, DSRC-based, safety application concepts capable of addressing them were developed. The crash imminent scenarios and the applications selected to be part of the VSC-A safety system is shown in Table 1. The VSC-A team together with the USDOT analyzed the scenarios in Table 1 and developed concepts for safety applications that could potentially address them through vehicle safety communications. This analysis resulted in the identification of the following safety applications as part of the VSC-A system:

Emergency Electronic Brake Lights (EEBL), defined as follows

The EEBL application enables a host vehicle (HV) to broadcast a self-generated emergency brake event to surrounding remote vehicles (RVs). Upon receiving the event information, the RV determines the relevance of the event and issues a warning to the driver, if appropriate. This application is particularly useful if the drivers' line of sight is obstructed by other vehicles or bad weather conditions (e.g., fog, heavy rain)

Forward Collision Warning (FCW), defined as follows

The FCW application is intended to warn the driver of the HV of an impending rear-end collision with an RV ahead in traffic in the same lane and direction of travel. FCW is intended to help drivers in avoiding or mitigating rear-end vehicle collisions in the forward path of travel.

Blind Spot Warning+Lane Change Warning (BSW +LCW), defined as follows

The BSW+LCW application is intended to warn the driver during a lane change attempt if the blind-spot zone into which the HV intends to switch is, or will soon be, occupied by another vehicle traveling in the same direction. Moreover, the application provides advisory information that is intended to inform the driver of the HV that a vehicle in an adjacent lane is positioned in a blind-spot zone of the HV when a lane change is not being attempted.

Do Not Pass Warning (DNPW), defined as follows

The DNPW application is intended to warn the driver of the HV during a passing maneuver attempt when a slower moving vehicle, ahead and in the same lane, cannot be safely passed using a passing zone which is occupied by vehicles in the opposite direction of travel. In addition, the application provides advisory information that is intended to inform the driver of the HV that the passing zone is occupied when a vehicle is ahead and in the same lane and a passing maneuver is not being attempted.

Intersection Movement Assist (IMA), defined as follows

The IMA application is intended to warn the driver of a HV when it is not safe to enter an intersection due to high collision probability with other RVs. Initially, IMA is intended to help drivers avoid or mitigate vehicle collisions at stop sign-controlled and uncontrolled intersections.

Control Loss Warning (CLW), defined as follows

The CLW application enables a HV to broadcast a self-generated, control-loss event to surrounding RVs. Upon receiving such event notification, the RV determines the relevance of the event and provides a warning to the driver, if appropriate.

Table 1 illustrates the mapping between the crash imminent scenarios and the safety applications defined above.

DEVELOPMENT OF THE TEST BED

Each OEM in the VSC-A Project developed a vehicle test bed to serve as a prototype platform for the V2V communications system. The OEMs jointly developed system specifications and performance requirements that served as the basis for the system and application developments. The test bed was based on a common prototype platform referred to as the On-Board Equipment (OBE). The selected OBE allowed development flexibility and was representative of current (or future) automotive grade processing power. The OBE contained a DSRC radio, a processor and various interfaces (e.g., for vehicle data, Global Positioning System (GPS) data, etc.). The test bed was an effective tool for validating safety application concepts, system test procedures and for answering critical research questions regarding V2V communications. Those issues included relative lane-level positioning, time synchronization, communications scalability and practical security and anonymity.

SOFTWARE ARCHITECTURE

In order to support the functionality of the safety applications described earlier and their development, the activities initially focused on the development of a system architecture based on various modules that could be upgraded independently from each other, if necessary. This approach allowed for fast and efficient prototyping throughout the development phase of the project. This architecture was used during the test bed design stage for the definition of the Hardware (HW) and Software (SW) architectures and required interfaces. The various modules forming the system test bed were categorized into the following major groups: Interface, Positioning & Security, Core, Safety Applications, Threat Process and Reporting, and Data Analysis. The system block diagram (Figure 1) shows the breakdown of the individual modules that make up each of the major module groupings. This provided a good framework for a comprehensive V2V safety system.

The focus of the system design activities was the core modules (Target Classification, Host Vehicle Path Prediction and Path History) and the positioning, security and safety application modules. The system design was based on the preliminary requirement specifications developed for each of the modules. Testing of the system resulted in updates to the modules throughout the project, culminating in the final test bed implementation. In the next section the software modules are described briefly.

SOFTWARE MODULES

The VSC-A software modules are composed of support and application functions. The support functions provide the interface to any external equipment and they calculate the necessary parameters to support the application modules and the Engineering Driver-Vehicle Interface (DVI). The primary software modules are:

- Threat Arbitration (TA)
- Driver-Vehicle Interface Notifier (DVIN)
- Target Classification (TC)
- Host Vehicle Path Prediction (HVPP)
- Path History (PH)
- Data Logger (DL)
- Engineering Graphical User Interface (EGUI)
- Sensor Data Handler (SDH)
- Wireless Message Handler (WMH)

The application modules evaluate potential safety threats based on the data and inputs from the support modules. The application modules contain the warning algorithms for the safety applications shown in Table 1. The SDH and WMH are basic, functional blocks necessary for parsing inputs from and submitting data to the software services of the system platform and those in use by the other support and application elements. The SDH interfaces to the vehicle Controller Area Network (CAN) gateway device to transmit and receive CAN messages and detect communication errors. It also connects to the GPS receiver to obtain National Marine Electronics Association (NMEA) data including Universal Coordinated Time (UTC) time, position, speed and heading, as well as raw GPS data. The SDH also interfaces to the external computing platform that executes the Real Time Kinematic (RTK) software to obtain accurate relative positioning of the neighboring vehicles. The WMH interfaces to the DSRC radio and to the Security Module (SM) software. It transmits and receives WAVE Safety Messages (WSM) using the SM to generate and verify message signatures. The TC categorizes the surrounding vehicles according to their position and heading relative to the HV, using the HVPP and the PH of the HV and RV. The TA arbitrates between multiple threats and chooses the one with the highest crash energy as the one to display to the driver and sends the respective request to the DVIN, which activates the corresponding alert in the EGUI.

The VSC-A team decided to use the shared memory interface concept. This allows for data in memory to be accessed by multiple modules for inter-process communication. This is advantageous, because there are many cases of one module supplying data to other functional blocks. For example, consecutive host and remote GPS time and position data points may be used by HVPP, PH, TC and the warning algorithms at the same time. The shared memory scheme used in the architecture fulfills the requirements for support of the VSC-A functionality while allowing for extensibility of the architecture.

ENGINEERING GUI

The EGUI is an "engineering-type" graphical user interface with the purpose to provide a simple engineering tool that could be used to understand, evaluate, and configure the VSC-A platform. It allows representation of visual and auditory vehicle driver warnings as a result of the application module processes. The touch-screen interface also allows the user to view and control parameters necessary for the operation of the VSC-A safety applications. Figure 2 shows examples of the graphical interface as depicted on a Video Graphics Array (VGA) touch screen.

This allowed the EGUI to display the warning states of a particular threat (e.g., DNPW in Figure 2). Only one of the warning screens is visible at any particular time. In order to ensure that the most important warning was shown on the DVI screen, the TA uses the threat level, relative speed, and location of the threat from each of the application modules to assess the severity and determine the highest priority request to be used by DVIN.

IN-VEHICLE HARDWARE INTEGRATION

The in-vehicle HW integration involved the selection, purchasing, installation and integration of all the HW and SW required for completion of the test bed. Table 2 identifies the model and manufacturer of the equipment installed on the VSC-A test bed vehicles.

MESSAGING STANDARDS

A major goal of the VSC-A Project was to define a single Over-the-Air (OTA) message whose contents could support all of the VSC-A safety applications as well as other safety applications that are likely to be developed in the future. That goal was achieved with the standardization of the SAE J2735 BSM [3]. An internal version of the OTA message was defined and implemented in the test bed with the objective testing verifying that this message supports all of the VSC-A applications. The BSM consists of Parts I and II. A proposal was prepared and presented for SAE to redefine both Parts I and II of the BSM. Part I consists of vehicle state data that is so critical for safety applications that it must be included in every BSM. Part II consists of data that is either required by applications at regular intervals (potentially at a reduced frequency), required to notify applications of a given event or optional for applications. Figure 3 shows the components and format of the BSM in SAE J2735.

The SAE J2735 conformant BSM uses the Distinguished Encoding Rules (DER) to encode the message for OTA transmission. In addition to the effort to develop and standardize the BSM, the VSC-A team also initiated a new SAE DSRC standards project (SAE J2945) for BSM minimum performance requirements. This standard will augment SAE J2735 to define rules necessary for effective V2V safety communications interoperability (e.g., minimum message rate, minimum data accuracy, etc.).

OBJECTIVE TESTING

OVERVIEW

The objective testing activity included the development of the Objective Test Procedures (OTPs) and test plan, conducting the objective tests, and analyzing the test results. The purpose of the objective testing was to ascertain that:

- The performance of the VSC-A system test bed was sufficient to enable the safety applications in the project
- The safety applications satisfied the minimum performance requirements developed in the system design activity of the project

The OTPs were developed for each application and were designed to include the most common scenarios that the application would encounter. The procedures included the following:

- True positive tests, where the objective is to get a warning
- False positive tests, where the objective is to suppress a warning because it is not needed

The outcomes of the objective tests were used by the Volpe National Transportation Systems Center (Volpe) to estimate the safety benefits opportunity for V2V communications based safety applications. In total, 33 test procedures were developed, 22 true positive tests and 11 false positive tests. For the benefits estimate, only the true positive tests which all had successful/unsuccessful criteria associated with them were evaluated. The OTPs were discussed with NHTSA and Volpe and agreed upon by all the participants. Following the OTP development, the test plan was written. It included the number of runs for each test, test speeds, validation criteria for each test (allowable speed ranges, etc.) and detailed setup procedures to make the OTPs as repeatable as possible. The test plan was also agreed upon by Volpe and NHTSA prior to the start of testing. The objective testing took place from June 1, 2009 to June 3 2009 at TRC in East Liberty, Ohio.

The data that was collected during the testing was recorded in a data logging and visualization tool called CANape [4]. CANape is a SW tool developed by Vector CANtech, Inc. and is used for the development, calibration and diagnostics of Electronic Control Units (ECUs) as well as data acquisition and analysis. The CANape software was customized by Vector for the VSC-A Project. Figure 4 shows an example of the primary screen that was used for the objective testing. The screen is divided into four quadrants as follows:

- Quadrant 1 contains a birds-eye view, which is a graphical representation of the location of the HV, centered at (0,0) and the RVs that the HV is in communication with

- Quadrant 2 contains the camera data which will consist of a single image, as shown below, or up to four images multiplexed together

- Quadrant 3 contains the HV's sensor data and GPS data

- Quadrant 4 contains the RV track data as determined by the TC core module

OBJECTIVE TEST RESULTS

The complete list of tests, the speed for the runs, the number of runs for each test and the test outcome is shown in Table 3. As can be seen from Table 3, all the applications passed the objective tests.

SUMMARY/CONCLUSIONS

The major accomplishments of the project are:

- Defined a set of high-priority, potential crash scenarios that could be addressed by V2V communication

- Selected and developed a set of V2V applications to address the above set of potential crash scenarios

- Defined efficient system architecture for V2V safety system where all VSC-A safety applications are enabled at the same time

- Successfully implemented a test bed with all the safety applications on a platform running an automotive grade processor (400 MHz)

- Successfully incorporated and evaluated in the test bed two relative positioning approaches (RTK and Single Point (SP))

- Successfully incorporated in the test bed the necessary OTA communication protocol (SAE J2735) and security protocol (IEEE 1609.2 Elliptic Curve Digital Signature Algorithm (ECDSA) [5] with Verify-on-Demand (VoD) [6])

- Defined OTPs for all the VSC-A safety applications, including true positive and false positive tests

- Successfully executed and passed all objective tests for all the VSC-A safety applications

• Refined, with field data, the required OTA message set for V2V safety (BSM within SAE J2735) which led to the recently published version of the standard [3]

• Conducted a study to quantify availability and accuracy of GPS-based relative positioning by using RTK and SP methods for V2V

• Confirmed that IEEE 1609.2 ECDSA with VoD functioned properly under all test conditions for the VSC-A safety applications

• Performed and analyzed initial scalability with up to 60 radios [8] to characterize channel behavior under IEEE 1609.4 [7] and under dedicated full time use of channel 172

Another outcome of the technical work was the identification of technical questions and topics that still need to be answered for any successful deployment:

• How does the system perform with large numbers of communicating nodes?

• How can security certificates be managed and privacy preserved?

• Are the standards sufficient for interoperability?

• What are requirements for data reliability and integrity?

• What are technical solutions for acceleration of market penetration?

• How to enhance the safety applications and system design?

• How to enhance relative vehicle positioning?

Those questions and topics are being addressed under the current NHTSA V2V safety roadmap [1] which outlines the next set of activities needed to support a NHTSA decision regarding V2V safety in 2013.

REFERENCES

1. DOT-Sponsored Research Activities: V2V Communications for Safety, http://www.intellidriveusa.org/research/v2v.php.

2. Vehicle Safety Communications - Applications (VSC-A) - First Annual Report, http://www.intellidriveusa.org/documents/2009/05/09042008-vsc-a-report.pdf

3. SAE International Surface Vehicle Standard, "Dedicated Short Range Communications (DSRC) Message Set Dictionary," SAE Standard J2735, Rev. Nov. 2009.

4. CANape, A Versatile Tool for Measurement, Calibration and Diagnostics of ECUs, Vector, http://www.vector.com/vi_canape_en.html?quickfinder=1.

5. IEEE Trial-use Standard 1609.2TM-2006, WAVE - Security Services for Applications and Management Messages, 2006.

6. Krishnan, H., Technical Disclosure, "Verify-on-Demand" - A Practical and Scalable Approach for Broadcast Authentication in Vehicle Safety Communication, IP.com number: IPCOM000175512D, IP.com Electronic Publication: October 10, 2008.

7. IEEE P1609.4TMD6.0, Draft Standard for Wireless Access in Vehicular Environments - Multi-channel Operation, IEEE Vehicular Technology Society, March 2010.

8. Ahmed-Zaid, F., Krishnan, H., Maile, M., Caminiti, L. et al., "Vehicle Safety Communications - Applications: Multiple On-Board Equipment (OBE) Testing," *SAE Int. J. Passeng. Cars – Mech. Syst.* **4**(1):547-561, 2011, doi: 10.4271/2011-01-0586.

CONTACT INFORMATION

Farid Ahmed-Zaid
Ford Motor Company
fahmedza@ford.com

ACKNOWLEDGMENTS

The CAMP VSC2 Participants would like to acknowledge the following USDOT personnel for their invaluable project support; Art Carter, Ray Resendes, and Mike Schagrin. The Participants would also like to thank VRTC personnel, especially Garrick Forkenbrock, for their outstanding support during the execution of the objective tests. Finally the Participants would like to express their appreciation to the following Volpe personnel; Wassim Najm, Bruce Wilson, and Jonathan Koopman for their support with the development and execution of the objective tests.

DEFINITIONS/ABBREVIATIONS

BSM
Basic Safety Message

BSW/LCW
Blind Spot Warning, Lane Change Warning

CAMP
Crash Avoidance Metrics Partnership

CAN
Controller Area Network

CLW
Control Loss Warning

DER
Distinguished Encoding Rules

DL
Data Logger

DNPW
Do Not Pass Warning

DSRC
Dedicated Short Range Communications

DVI
Driver-Vehicle Interface

DVIN
Driver-Vehicle Interface Notifier

ECDSA
Elliptic Curve Digital Signature Algorithm

ECU
Electronic Control Unit

EEBL
Emergency Electronic Brake Lights

EGUI
Engineering Graphical User Interface

FCW
Forward Collision Warning

GES
General Estimated Systems

GPS
Global Positioning System

HV
Host Vehicle

HVPP
Host Vehicle Path Prediction

HW
Hardware

IMA
Intersection Movement Assist

ITS
Intelligent Transport Systems

NHTSA
National Highway Traffic Safety Administration

NMEA
National Maritime Electronics Association

OBE
On-Board Equipment

OEM
Original Equipment Manufacturer

OTA
Over-the-Air

OTP
Objective Test Procedure

PH
Path History

RTK
Real-Time Kinematic

RV
Remote Vehicle

SDH
Sensor Data Handler

SM
Security Module

SP
Single Point (positioning)

SW
Software

TA
Threat Arbitration

TC
Target Classification

TRC
: Transportation Research Center

USDOT
: United States Department of Transportation

UTC
: Universal Coordinated Time

V2V
: Vehicle-to-Vehicle

VGA
: Video Graphics Array

VoD
: Verify-on-Demand

VRTC
: Vehicle Research and Test Center

VSC2
: Vehicle Safety Communications 2 (Consortium)

VSC-A
: Vehicle Safety Communications - Applications

WMH
: Wireless Message Handler

WSM
: Wave Short Message

DISCLAIMER

This material is based upon work supported by the National Highway Traffic Safety Administration under Cooperative Agreement No. DTNH22-05-H-01277. Any opinions, findings, and conclusions or recommendations expressed in this publication are those of the Author(s) and do not necessarily reflect the view of the National Highway Traffic Safety Administration.

APPENDIX A

OBJECTIVE TEST PROCEDURE EXAMPLE AND TEST RESULTS

In this appendix we provide an example of the test plan and OTP together with the results of the testing. The chosen example is the FCW, Test 1.

FCW OBJECTIVE TEST PROCEDURES

FCW is a V2V, communication-based, safety feature that issues a warning to the driver of the HV in case of an impending rear-end collision with a vehicle ahead in traffic in the same lane and direction of travel. FCW is designed to help drivers in avoiding or mitigating rear-end vehicle collisions in the forward path of travel.

FCW-T1: HV Travel at a Constant Speed to a Stopped RV

Background

This test begins with the HV traveling on a straight, flat road at 50 mph. Ahead of the HV, in the same lane, is a single RV stopped in the lane of travel. The test determines whether the countermeasure's required collision alert occurs at the expected range. This test especially explores the ability of the countermeasure to accurately identify stationary in-path targets on a flat, straight road.

Test Setup

Figure 5 shows the vehicle positions and test setup for Test 1.

Cones with flags are placed so the driver of the HV is aware of the vehicle's location in reference to the required maneuvers. These flags are located by their distance from the starting point for the HV. It is assumed that flags will be placed using an accurate GPS handheld receiver. Alternate methods of flag location can be used. Flag locations are:

- A red flag is placed at the starting point where the HV begins its maneuver (cone not shown)
- A yellow flag is placed at the point where the HV reaches the target speed (cone HV-A), at least 650 meters from the red flag
- A white flag is placed at the earliest valid (from the driver's perspective) WARN point (cone HV-B)

A checkered flag is placed where the HV will make an evasive maneuver by changing lanes if the WARN has failed to occur (cone HV-C) which is positioned at 90 percent of the allowable alert range. At the test speed of 50 mph, this is 9 meters from HV-B cone

A green flag is placed at the stopping position for the RV (cone RV-A), at least 800 meters from the red flag

Driving Instructions

- The RV begins at the starting point and stops with its front bumper at the green flag
- The HV starts accelerating at least 800 meters behind the RV in the same lane to reach a speed of 50 mph
- The HV Cruise Control is set at the required speed of 50 mph
- The HV Cruise Control shall be engaged at least 150 meters behind the RV
- The warning will be given at around the nominal warn range (cone HV-B) after which the HV will change lane

[Note: If the warning is not given when the HV reaches the checkered flag (cone HV-C), the HV shall make an evasive maneuver by changing lanes and come to a safe stop in the adjacent lane.]

Successful Criteria

- The collision alert shall occur within the ranges specified in Table 4 in order to pass the run
- If at least six runs out of eight runs pass, then the test is successful

Unsuccessful Criteria

- A run is unsuccessful if any of the conditions below occur:
    - Collision alert occurrence outside the range calculated in Table 4 using run-specific variables
    - The warning is missed such that the HV passes cone HV-C and no alert is triggered
    - If at least three runs out of eight runs fail, the test is unsuccessful

Table 4. Alert Range for FCW Test 1

	Collision Alert Test
Maximum Range	93.7
Nominal Range	85.2
Minimum Range	76.7

Evaluation Criteria

Number of Valid Test Runs	HV Speed (mph)	RV Speed (mph)	Number of Successful Test Runs
8	50	0	≥ 6

FCW OBJECTEIVE TEST 1 RESULTS

For the FCW application to pass, the warning had to come between the maximum and minimum alert range that was calculated for each run. As can be seen from the test results table (Table 5), the application was successful in all the runs for the test.

APPENDIX B

TABLES AND FIGURES

Table 1. Mapping of VSC-A Program Applications to Crash Imminent Scenarios

	V2V Safety Applications/ Crash Scenarios	EEBL	FCW	BSW	LCW	DNPW	IMA	CLW
1	Lead Vehicle Stopped		✓					
2	Control Loss without Prior Vehicle Action							✓
3	Vehicle(s) Turning at Non-Signalized Junctions						✓	
4	Straight Crossing Paths at Non-Signalized Junctions						✓	
5	Lead Vehicle Decelerating	✓	✓					
6	Vehicle(s) Not Making a Maneuver – Opposite Direction					✓		
7	Vehicle(s) Changing Lanes – Same Direction			✓	✓			

Table 2. VSCA Test Bed Hardware List

Item Description	Manufacturer	Model
GPS Receiver	NovAtel®	OEMV® Flexpak V1-RT20A
GPS Antenna	NovAtel®	GPS-701-GG
LCD VGA Monitor	Xenarc	700TSV-B
USB CCD Monochrome Camera	The Imaging Source	DMK 21BU04
Car PC	Logic Supply	Voom PC-2
Inertial Measurement Unit	Silicon Sensing	DMU
OBE Vehicle CAN interface	Smart Engineering Tools	Netway 6
DSRC Antenna	Nippon Antenna	DEN-HA001-001
Ethernet Switch	Netgear	GS105

Table 3. Objective Test Scenarios and Results

Test Scenario	Description	Speeds	Number of Runs	Type of Test	Result
EEBL-T1	HV at constant speed with decelerating RV in same lane	50	8	True Positive	Successful
EEBL-T2	HV at constant speed with decelerating RV in left lane on curve	50	8	True Positive	Successful
EEBL-T3	HV at constant speed with decelerating RV in same lane and obstructing vehicle in between	50	8	True Positive	Successful
EEBL-T4	HV at constant speed with mild-decelerating RV in same lane	50	2	False Positive	N/A

Test Scenario	Description	Speeds	Number of Runs	Type of Test	Result
EEBL-T5	HV at constant speed with decelerating RV in 2nd right lane	50	2	False Positive	N/A
FCW-T1	HV travel at a constant speed\RV stopped	50	10	True Positive	Successful
FCW-T2	HV travel behind RV1\RV1 travel behind RV2\RV2 stopped	50	10	True Positive	Successful
FCW-T3	HV drive on a curve\RV stopped at the curve	50	8	True Positive	Successful
FCW-T4	HV tailgate RV	50	2	False Positive	N/A
FCW-T5	HV follows RV\RV brakes hard	40	10	True Positive	Successful
FCW-T6	HV driving into a curved right lane\RV stopped in the left curved lane	50	2	False Positive	N/A
FCW-T7	HV travels behind a slower RV	50	10	True Positive	Successful
FCW-T8	HV changes lanes behind a stopped RV	50	8	True Positive	Successful
FCW-T9	HV approaches two RVs in left and right adjacent lanes and passes between them	50	2	False Positive	N/A
BSW/LCW-T1	LCW Warning, Left	50	8	True Positive	Successful
BSW/LCW-T2	LCW Warning, Right	50	8	True Positive	Successful
BSW/LCW-T3	LCW Warning, Right with Left BSW Advisory	50	9	True Positive	Successful
BSW/LCW-T4	BSW Advisory Alert, Left	50	8	True Positive	Successful
BSW/LCW-T5	BSW Advisory Alert, Right	50	8	True Positive	Successful
BSW/LCW-T6	No Warning or Advisory for RV behind	50	2	False Positive	N/A

Test Scenario	Description	Speeds	Number of Runs	Type of Test	Result
BSW/LCW-T7	No Warning or Advisory for RV far Right	50	2	False Positive	N/A
BSW/LCW T8	LCW Warning in Curve, Right	35	8	True Positive	Successful
DNPW-T1	Attempt to pass with oncoming RV in adjacent lane	25/35	10	True Positive	Successful
DNPW-T2	Attempt to pass with stopped RV in adjacent lane	30/40	10	True Positive	Successful
DNPW-T3	Attempt to pass with oncoming RV not in adjacent lane	45	2	False Positive	N/A
IMA-T1	Variable speed approaches with stopped HV/moving RV/open intersection	20/30/40/50	12	True Positive	Successful
IMA-T2	Stopped HV/moving RV/open intersection	35/50	4	False Positive	N/A
IMA-T3	Variable speed approaches with moving HV/moving RV/open intersection	15/25/35/45	16	True Positive	Successful
IMA-T4	Moving HV/moving RV/open intersection	25	4	False Positive	N/A
IMA-T5	Stopped HV/moving RV/open intersection/parked vehicle	20/30/40/50	12	True Positive	Successful
CLW-T1	HV at constant speed with CLW RV in same lane ahead in same travel direction	40	8	True Positive	Successful
CLW-T2	HV at constant speed with CLW RV in 2nd right lane	30	2	False Positive	N/A
CLW-T3	HV at constant speed with CLW RV in adjacent lane ahead in opposite travel direction	30	12	True Positive	Successful

Table 5. FCW Test 1 Results

	Actual Values at Alert Onset				Calculated Run-Specific Alert Ranges					
Run	**HV Speed**	**HV Accel**	**RV Speed**	**RV Accel**	**Maximum (meters)**	**Nominal (meters)**	**Minimum (meters)**	**Actual Alert Range (meters)**	**Pass/Fail**	**Headway**
FCW T1	50 mph 22.35 m/s	0 g	0 mph	0 g	93.7	85.2	76.7			6.71
1	49.70 mph 22.21 m/s	0 g	0 mph	0 g	92.8	84.4	75.9	84	Pass	6.80
2	50.0 mph 22.35 m/s	0 g	0 mph	0 g	93.7	85.2	77.7	86	Pass	6.83
3	49.68 mph 22.21 m/s	0 g	0 mph	0 g	91.7	83.4	75.0	85	Pass	6.76
4	50.31 mph 22.49 m/s	0 g	0 mph	0 g	94.1	85.5	77.0	85	Pass	6.80
5	49.69 mph 22.21 m/s	0 g	0 mph	0 g	92.8	84.4	75.9	85	Pass	6.80
6	49.50 mph 22.13 m/s	0 g	0 mph	0 g	93.1	84.6	76.2	80	Pass (A # of communication outages from the RV)	6.82
7	49.84 mph 22.28	0 g	0 mph	0 g	93.9	85.4	76.8	83	Pass	6.73

	Actual Values at Alert Onset				Calculated Run-Specific Alert Ranges					
Run	**HV Speed**	**HV Accel**	**RV Speed**	**RV Accel**	**Maximum (meters)**	**Nominal (meters)**	**Minimum (meters)**	**Actual Alert Range (meters)**	**Pass/Fail**	**Headway**
	m/s									
8	49.75 mph 22.24 m/s	0 g	0 mph	0 g	92.7	84.3	75.9	82	Pass (A # of communication outages from the RV)	6.79
9	50.35 mph 22.51 m/s	0 g	0 mph	0 g	93.2	84.7	76.3	85	Pass	6.66
10	49.71 mph 22.22 m/s	0 g	0 mph	0 g	91.7	83.3	75.0	84	Pass	6.75

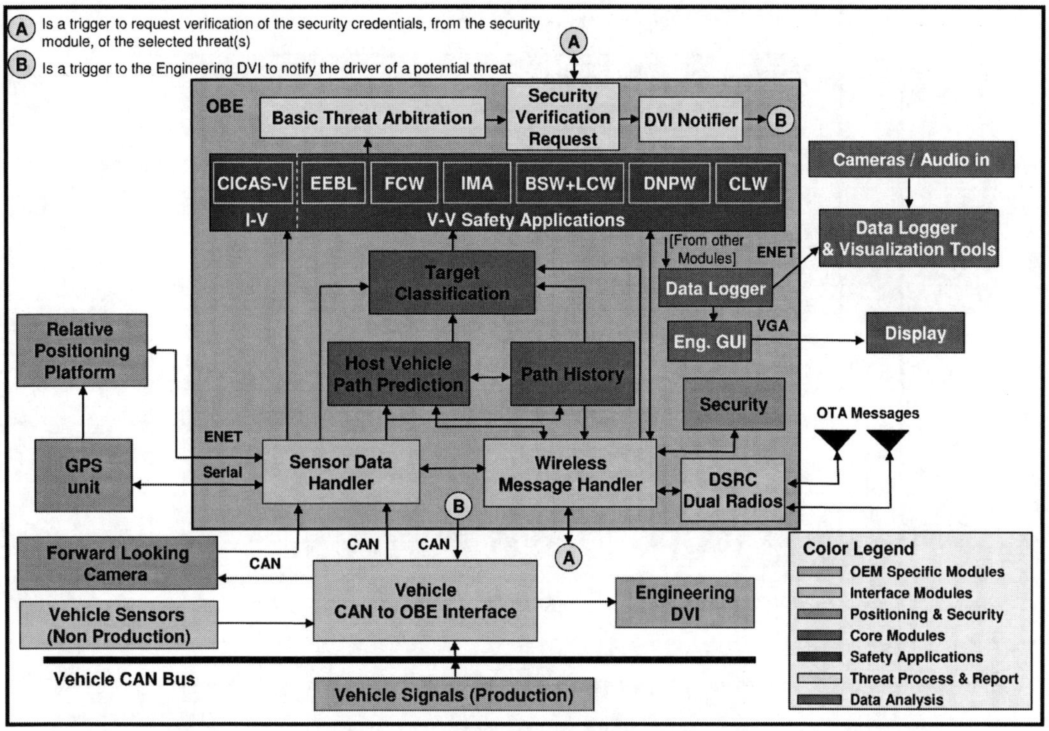

Figure 1. VSC-A System Block Diagram

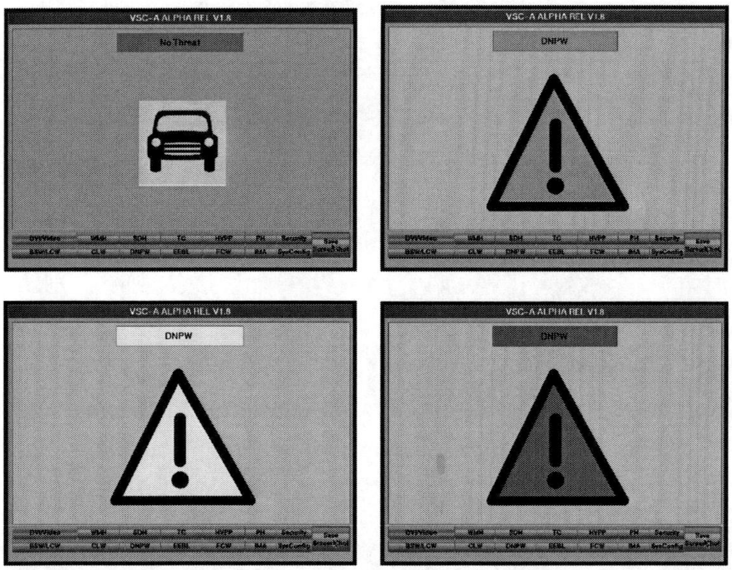

Figure 2. DVIN Stages (left to right, top to bottom) No Threat, Threat Detected, Inform Driver, Warn Driver

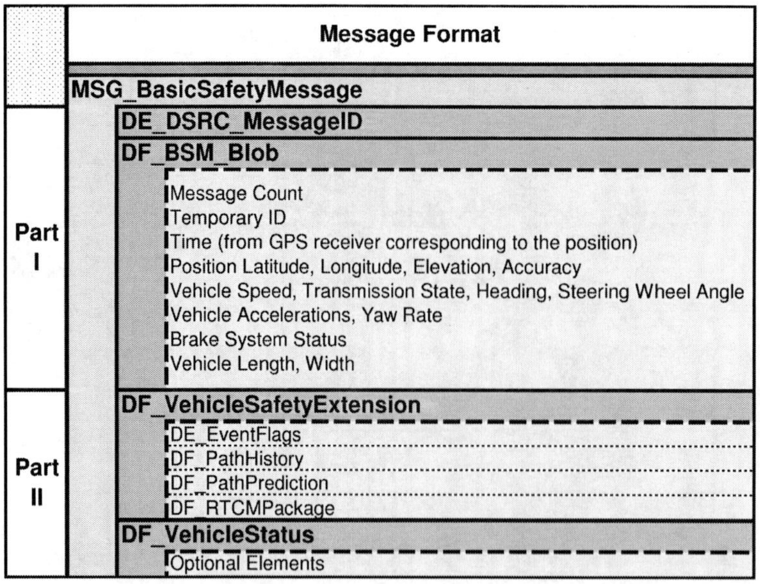

Figure 3. SAE J2735 Rev 35 Basic Safety Message Format

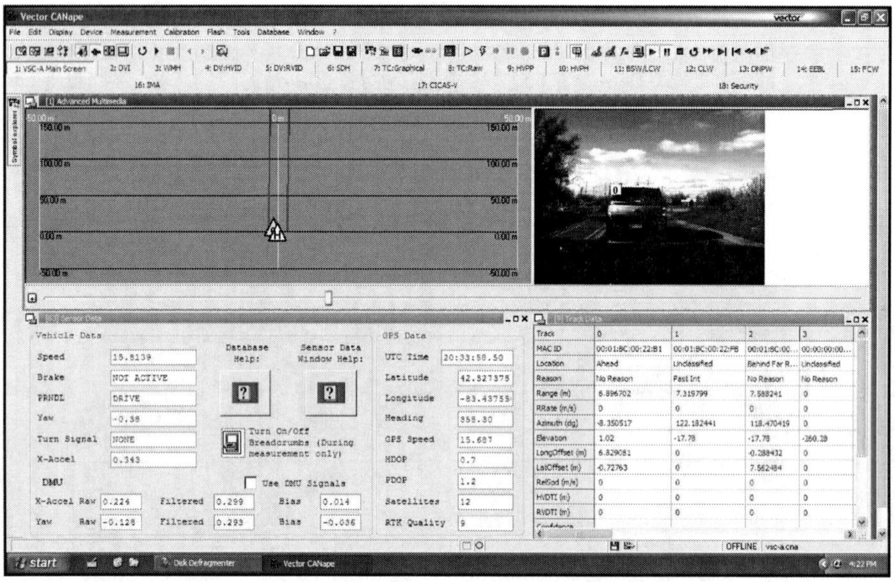

Figure 4. Example Layout Screen for OTP Testing

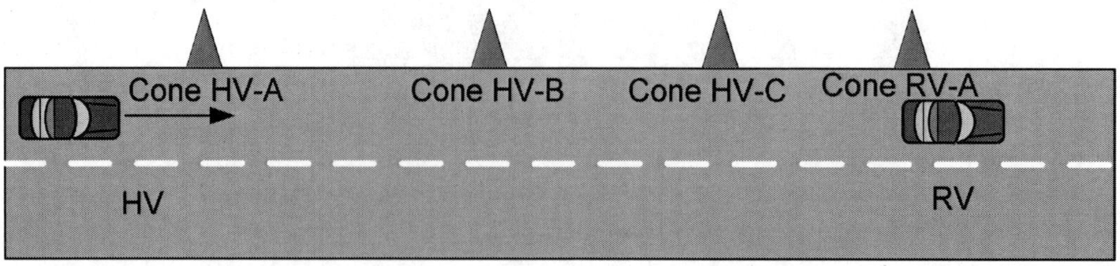

Figure 5. FCW Test 1 Test Setup - RV in Same Lane

SAE International

Vehicular Networks for Collision Avoidance at Intersections

2011-01-0573
Published
04/12/2011

Seyed Reza Azimi, Gaurav Bhatia and Ragunathan (Raj) Rajkumar
Carnegie Mellon University

Priyantha Mudalige
GM Technical Center

doi:10.4271/2011-01-0573

ABSTRACT

A substantial fraction of automotive collisions occur at intersections. Statistics collected by the Federal Highway Administration (FHWA) show that more than 2.8 million intersection-related crashes occur in the United States each year, with such crashes constituting more than 44 percent of all reported crashes [12]. In addition, there is a desire to increase throughput at intersections by reducing the delay introduced by stop signs and traffic signals. In the future, when dealing with autonomous vehicles, some form of co-operative driving is also necessary at intersections to address safety and throughput concerns.

In this paper, we investigate the use of vehicle-to-vehicle (V2V) communications to enable the navigation of traffic intersections, to mitigate collision risks, and to increase intersection throughput significantly. Specifically, we design a vehicular network protocol that integrates with mobile wireless radio communication standards such as Dedicated Short Range Communications (DSRC) and Wireless Access in a Vehicular Environment (WAVE). This protocol relies primarily on using V2V communications, GPS and other automotive sensors to safely navigate intersections and also to enable autonomous vehicle control. Vehicles use DSRC/WAVE wireless media to periodically broadcast their position information along with the driving intentions as they approach intersections. We used the hybrid simulator called GrooveNet [1, 2] in order to study different driving scenarios at intersections using simulated vehicles interacting with each other. Our simulation results indicate that very reasonable improvements in safe throughput are possible across many practical traffic scenarios.

INTRODUCTION

Current human driver-based intersections which are managed by stop signs and traffic lights are not entirely safe, based on Federal Highway Administration (FHWA) statistics [12]. Our goal is to design new methods to manage intersections, which lead to fewer collisions and less travel delay for vehicles crossing at intersections. Various driverless vehicles have been developed and tested at intersections, such as in the DARPA Urban Challenge [3] and General Motor's EN-V, which has been recently unveiled in Shanghai, China [4]. Our focus is to use vehicle-to-vehicle (V2V) communication as a part of co-operative driving in the context of autonomous vehicles to manage intersection traffic efficiently and safely.

Past work in this domain includes the use of Vehicle to Infrastructure (V2I) communication by having a centralized system in which all vehicles approaching an intersection communicate with the intersection manager. The intersection manager is the computational infrastructure installed at intersections and to make reservations for each approaching vehicle and manages all vehicles crossing the intersection [5,6,7,14,15,18].Installing centralized infrastructure at every intersection is not very practical due to prohibitively high total system costs. In this work, we advocate the use of Vehicle-to-Vehicle (V2V) communications and a distributed intersection algorithm that runs in each vehicle. Our focus in this paper is on (a) designing new protocols for V2V based-intersection management, (b) extending an advanced mobility simulator for vehicles, and (c) comparing our protocols to the operational efficiency of conventional intersections with stop signs and traffic lights.

The rest of this paper is organized as follows. Section 2 introduces the collision-detection algorithm used in our proposed intersection protocols. Section 3 contains intersection protocols used to manage various intersection scenarios. This section consists of a stop-sign model, a traffic light model and three V2V-based protocols: V2V Stop-Sign Protocol (SSP), Throughput-Enhancement Protocol (TEP) and Throughput-Enhancement Protocol with Agreement (TEPA). In Section 4, we describe the implementation of our protocols using the GrooveNet hybrid simulator, with new mobility and trip models. Section 5 contains the evaluation of our intersection protocols. Section 6 presents our concluding remarks.

COLLISION DETECTION AT INTERSECTIONS

We currently define an intersection as a perfect square box which has predefined entry and exit points for each lane connected to it. The trajectory of the vehicle crossing the intersection, is supposed to be the path taken by the vehicle from the entry to the exit point. We assume that each vehicle has access to a map database that provides routing, lane and road information, in which each segment of the road has a unique identifier (ID). Intersections are also identified by unique IDs in this map database.

Suppose *Arrival-Time* is the time at which a vehicle arrives at an entrance of the intersection and *Exit-Time* is the time at which the vehicle exits the intersection area. We refer to the part of the road that a vehicle is currently on as its **current road segment (CRS)**, and the part of the road that the vehicle will be moving to after the current road segment as the **next road segment (NRS)**. In the context of an intersection, CRS corresponds to the road segment that a vehicle is on before the intersection, and NRS represents the road segment that the vehicle will be on after crossing the intersection.

Each vehicle broadcasts CRS, NRS, current lane number, as well as the Arrival-Time and the Exit-Time, to all the other vehicles in its communication range. Vehicles are also assumed to have access to a global positioning system (GPS) with locally generated Radio Technical Commission for Maritime (RTCM-104) corrections to achieve Real-time Kinematic (RTK) solution.

Vehicles use this information to determine the other vehicles' turn types. Figure 1 shows an example of this, wherein a vehicle intends on entering the intersection from the east and exiting to the south. Based on the CRS, NRS and lane number, we can figure out that the vehicle is going to make a right turn. We assume in this paper that vehicles can make different turns regardless of their current lane number but they should stay in the same lane after passing any intersection and do not switch lanes. It is relatively easy to restrict this behavior, assumed for convenience here.

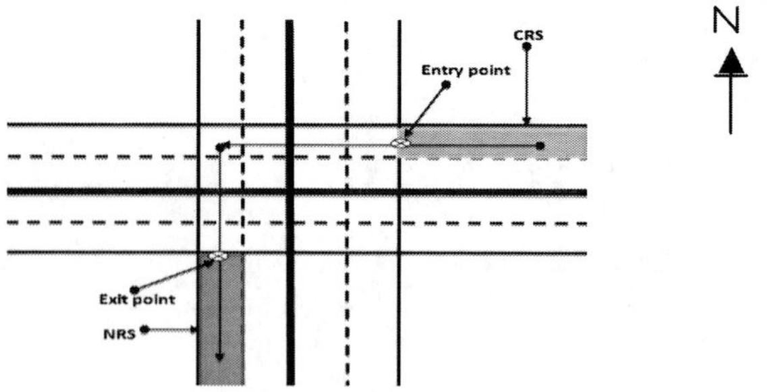

Figure 1.

We first identify the conditions required for two or more vehicles to collide at an intersection.

If a vehicle enters an intersection while another vehicle is in the intersection area, their (*Arrival-Time*, *Exit-Time*) intervals must overlap. Two vehicles being inside the same intersection at the same time is a necessary, but not sufficient condition for a collision. In Figure 2 (a), two vehicles are within the intersection at the same time but not occupying the same space. Figure 2(b) shows a scenario in which a vehicle is coming from the south and turning right while the other vehicle is coming from the north and also turning to its right. In this case, both vehicles can cross the intersection at the same time without a collision.

A collision occurs if the following conditions are all true:

1. **Same Intersection**: vehicles are at the same intersection.

2. **Time Conflict**: vehicles have overlapping (Arrival-Time, Exit-Time) intervals.

3. **Space Conflict**: vehicles occupy the same space while crossing the intersection.

If any of the above three conditions is false, then there will be no collision and vehicles can safely continue along their trajectory.

Our **Collision Detection Algorithm for Intersections** (CDAI) will be run on each vehicle that crosses a transaction, with information exchanged among vehicles approaching, crossing and leaving the intersection. The algorithm uses path prediction to determine any space conflicts with other vehicles trying to cross the intersection. Each lane on the road is considered to be a polygon, which starts from the previous intersection and ends at the next approaching intersection. Then, CDAI predicts the space (or region) which will be occupied by the vehicle during its trajectory. Utilizing the CRS (current road segment), current lane, and NRS (next road segment) information for each vehicle, CDAI predicts the path taken by the vehicle to cross the intersection and generates two polygons: the first polygon is related to the vehicle's CRS and current lane, and the second polygon is

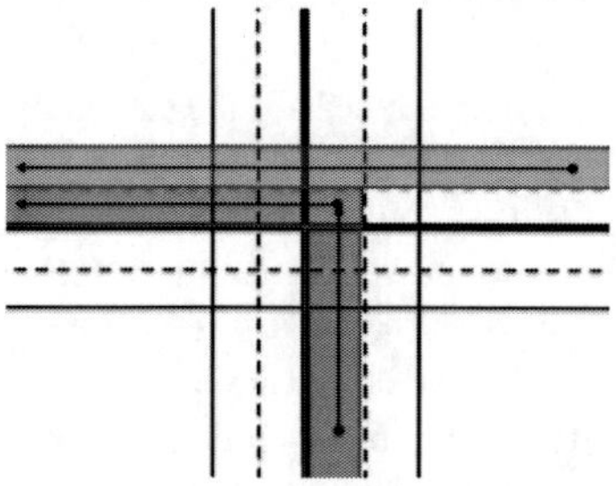

Figure 2(a)

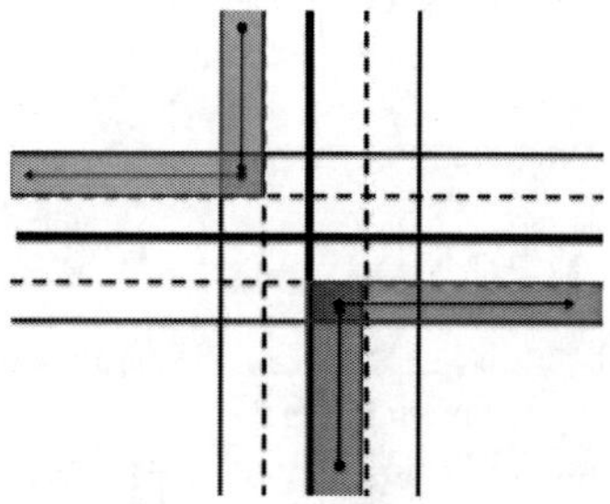

Figure 2(b)

Example scenarios in which no space conflict occurs at the intersection

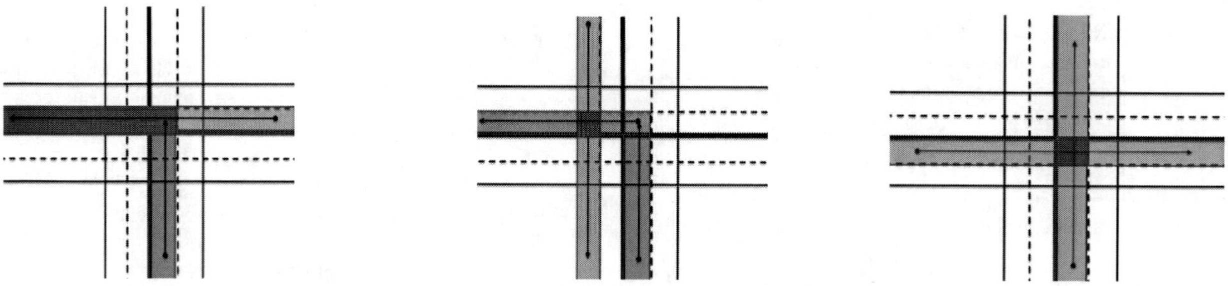

Figure 3. Three example scenarios of space conflict

related to the vehicle's NRS. Each polygon's height is the length of the road between two consecutive intersections and the polygon's width is the lane width. So, for each vehicle, these two polygons together form the complete spatial region related to its path, which we refer to as its **Trajectory Box (TB)**. As illustrated in Figure 3, a collision can potentially occur if two vehicles approaching the same intersection have intersecting "TBs".

To find out if the TBs of two vehicles intersect, we use the *Separating Axis Theorem* [8, 9]. The Separating Axis Theorem states that, for a pair of convex polygons that are not in a state of collision, there exists an axis perpendicular to an edge of one of the polygons that has no overlap between the projected vertices of the two polygons. This theorem can be simplified for our purposes since we are only dealing with two-dimensional rotated rectangles. Therefore, each polygon is tested against the four axes of the other polygon and if all projections overlap, a collision is detected. An optimization on this theorem exists for two-dimensional rotated rectangles, wherein, the polygon-under-test is rotated and centered on the intersection of the *x*-axis and the *y-axis*, and hence projections need to occur for only 2 axes [13].This solution works for any collision possibility, even for *cross-collisions* where a collision occurs between two polygons perpendicular to each other.

If a potential collision is detected by CDAI, it uses a **priority-based policy** to assign priorities to vehicles so that an unambiguous and repeatable precedence order in which vehicles cross the intersection can be established. For prioritizing the movement of vehicles at the intersection, the "first come, first served" (FCFS) algorithm is used. Based on FCFS, the first car arriving at the intersection is the first one crossing and leaving the intersection. Even though FCFS is an efficient algorithm, it can lead to a deadlock situation in particular scenarios such as when several vehicles get to an intersection at the same time (or very close to each other). To avoid any deadlocks, three **tie-breaking policies** are applied for vehicles with the same arrival time in the following sequence:

1. Roads are categorized as *primary roads* and *secondary roads* based on the roadmap database information, and higher priority is assigned to vehicles arriving at an intersection using a primary road than vehicles arriving using a secondary road.

2. If vehicles arrive at an intersection at the same time and using the same type of (primary or secondary) road, priorities are assigned based on their trajectories and whether turns are required. Specifically, vehicles going straight have higher priority than vehicles turning right, with vehicles turning left getting the lowest priority.

3. If all the previous conditions still result in a tie among two or more vehicles, the *Vehicle ID* (VID) which is unique for each vehicle is used to break ties - the vehicle holding a higher VID is given higher priority to cross the intersection.

The CDAI decision is made within each vehicle based on the information communicated using V2V.The algorithm will alert a vehicle if it can cross the intersection safely or, if any collisions are predicted, the vehicle must stop.

INTERSECTION PROTOCOLS

In this section, we describe three protocols[1]: the Stop-Sign Protocol (SSP), the Throughput-Enhancement Protocol (TEP) and Throughput-Enhancement Protocol with Agreement (TEAP). We will specify their functionality under various scenarios. The contents of messages communicated among the vehicles will be detailed in the next section. In the following three protocols, we assume that all vehicles have the same shape and physical dimensions. They do not use any controller model, which means that there is no consideration about a vehicle's movement during acceleration and deceleration. The communication medium has been assumed to be perfect; therefore, no packet loss occurs.

STOP-SIGN PROTOCOL (SSP)

In this protocol, we assume that stop-signs are not physically present at the intersection but vehicles obey the stop-sign rules when they approach an intersection. Vehicles only use V2V communications. Let t_s be the minimum amount of time in seconds that a car must wait at an intersection before proceeding. When vehicles approach an intersection, they must obey the rules of a stop sign which is to waitt_s seconds even if there is no other vehicle around. The FCFS priority policy mentioned in the previous section is obeyed by each vehicle. Vehicles also use STOP and CLEAR safety messages at the intersection in order to inform other vehicles in range about their current situation and movement parameters. The following rules are applicable.

• **Sending STOP:** As a vehicle approaches an intersection, it transmits a STOP safety message. Any vehicles within range will receive that message. When the vehicle arrives at the intersection, it also comes to a complete stop for t_s seconds.

• **On Receiving STOP:** On receipt of a STOP message, a vehicle uses the CDAI scheme described earlier, except for the *Space Conflict* rule. If more than one car arrives at the same intersection and will be inside the intersection area for an overlapping interval of time, priorities will get assigned to them and the vehicle with the highest priority will cross the intersection aftert_s seconds pass. Lower-priority vehicles will remain stopped even after t_s seconds, waiting to receive a CLEAR message.

• **Sending CLEAR:** When a vehicle crosses the intersection secondary and travels a distance defined by a threshold parameter D_{TC}, it broadcastsCLEAR messages indicating that the intersection is now safe to pass.

• **On Receiving CLEAR:** On receiving this message, the vehicle checks if it has stopped for at leastt_s seconds and, if true, it then checks if the sender of the CLEAR message is the same as the sender of the STOP message. The FCFS, priority and tie-breaking rules are again applied. Ift_s seconds have not passed as yet, the vehicle remains stopped while processing received messages to make a decision when the t_s seconds ends.If several vehicles are stopped at the intersection, by re-applying the priority policy, each vehicle decides if it should remain stopped or it can cross the intersection next as it has the highest priority among all stopped vehicles at the intersection.

THROUGHPUT ENHANCEMENT PROTOCOL (TEP)

This protocol is designed to manage intersection crossings by pure V2V communication without using any infrastructure such as stop-signs, traffic lights, sensors and cameras. The goal is to enhance the throughput at intersections without causing collisions. Vehicles again use STOP and CLEAR safety messages to interact with other vehicles. We define the throughput of an intersection based on the delay of all vehicles trying to cross the intersection. The following rules are applicable to each vehicle.

• **Sending STOP:** Every vehicle has access to its own GPS coordinates, speed and also to the map database; using these values, it computes the distance to the approaching intersection. The accuracy of this distance prediction is directly related to GPS accuracy. If the current distance of the host vehicle from the other vehicle is not greater than a threshold parameter D_{STOP}, then it starts sending periodic STOP messages (with the goal of informing other vehicles within range that it is getting close to the intersection). The STOP message will be sent with frequencyf_{STOP}.

• **Sending CLEAR:** When the vehicle exits the intersection, it sends periodic CLEAR messages with frequencyf_{CLEAR} until it travels further than a threshold value D_{CLEAR} from the exit point of the intersection. This behavior lets other vehicles know that the intersection is no longer in use by this vehicle.

• **On Receiving STOP:** On receiving a STOP message, the vehicle checks if all three collision conditions are satisfied. If even one of the conditions is not satisfied, then it means that the vehicle can cross the intersection without a collision with the sender of the STOP message. Otherwise, the vehicle acts based on the priority assigned to it using the priority policy. If it has lower priority than the sender of the STOP message, it comes to a complete stop at the intersection. Else, it has higher priority and ignores this message. In the latter case, the vehicle will have precedence at the intersection. Note that a

[1] Our protocols are inspired at least in part by Kurt Dresner's work [11]. Our focus is exclusively on V2V-based protocols, and our contributions include support for intersection management protocols in GrooveNet [1, 2], detailed evaluations and ongoing implementations in real vehicles.

vehicle which first started sending STOP messages may be superseded by a later vehicle due to priority considerations.

• **On Receiving CLEAR:** Each vehicle stores the information within received STOP messages which made it stop at the intersection. On receiving a CLEAR message, the vehicle checks if this message is sent from the sender of the last STOP message that has higher priority and because of which the vehicle is waiting at the intersection. This check is possible by just looking at the unique VID embedded in the message. If the VID of the CLEAR message is the same as the VID of the last processed STOP message, then the space that the vehicle needs to occupy for crossing the intersection is now clear.

Using TEP, vehicles stop at the intersection only if the collision detection algorithm predicts a collision and assigns a lower priority to them based on the messages it receives from all vehicles at the intersection. If no collision potential is detected or the highest priority is determined among contending vehicles, a vehicle can ignore other STOP messages, broadcast its own STOP messages to notify other vehicles, and cross the intersection safely. Multiple vehicles can be inside the intersection area at the same time if no space conflict occurs based on the collision detection policy's results. These rules increase the throughput of the intersection by decreasing the average delay time relative to the situation that vehicles should stop at the intersection. (We are currently studying enhancements to this protocol which will enable vehicles to slow down instead of coming to a complete stop when there are vehicles with higher priority entering the intersection. Evaluations of this scheme will be reported in the near future).

A reader might note that TEP implicitly assumes that V2V messages are not lost. While TEP will indeed work better with a very reliable wireless medium, the periodic transmission of STOP and CLEAR messages is targeted at a lossy communications medium and the protocol can tolerate *some* lost messages.

THROUGHPUT ENHANCEMENT PROTOCOL WITH AGREEMENT (TEPA)

This protocol is built on TEP and is explicitly designed to handle lost V2V messages. Additional CONFIRM and DENY messages are used to perform explicit handshaking between vehicles approaching the same intersection. Each vehicle makes its own local decision as in the previous protocols, but each vehicle announces its decision to cross the intersection by sending a CONFIRM or DENY message to either adhere to or override a decision made by another vehicle. On receiving a STOP message from another vehicle, the receiver will also send a message to acknowledge the reception of the message. The following rules are used by each vehicle *in addition to* the rules used by TEP:

• **Sending CONFIRM:** if no collision with the sender of a STOP message is predicted by CDAI, this message is sent first. It is also sent if a collision is predicted and a lower priority is assigned to the receiver of the STOP message. In this case, the receiver of the STOP message comes to a complete stop and waits for a CLEAR message.

• **Sending DENY:** If a collision is predicted and the receiver of the STOP message has a higher priority than its sender, the vehicle will send a DENY message to inform the sender of the STOP message that the latter's decision has been overridden and that this vehicle will *not* stop at the intersection.

• **On Receiving CONFIRM:** if the vehicle had sent a STOP message earlier, it has higher priority than the sender of the CONFIRM message and continues to proceed with its current decision.

• **On Receiving DENY:** if the vehicle had sent a STOP message later, it now has lower priority than the sender of the DENY message and must wait for a CLEAR message when it must re-evaluate the situation.

The collision detection scheme used in our intersection protocols ensures that two vehicles will not occupy the same space at the same time while crossing the intersection. Essentially, if there is any trajectory conflict, then one of the cars will be assigned a higher priority based on the priority policy, and the other one will wait for a CLEAR message without entering the intersection area. This prevents any collision between vehicles crossing the intersection.

DISTANCE KEEPING

In order to ensure a safe distance between cars, a distance-keeping protocol known as the *Car-Following Model* is used. This model is designed to control the mobility of vehicles while moving towards and exiting the intersection. A message of type *Generic* is sent at a regular interval and contains information about a vehicle's position, current lane, as well as current and projected map DB locations. On receiving this message, each vehicle checks if it is on the same road segment and the same lane as the sender. If this is the case, then by comparing its current GPS position with the sender's position, the vehicle determines if the sender is in front or behind it. In case of being behind the sender's vehicle, the vehicle adjusts its current velocity to the speed of the vehicle in front to prevent any collision. The vehicle does not need to have the same speed as the leader vehicle unless the distance between them is less than a threshold D_{follow}. Otherwise, it can maintain its current velocity which is related to the road's speed limit.

INTERSECTION SAFETY MESSAGE TYPES

We now describe in detail the content of transmitted messages.

The STOP message contains 9 parameters:

• **Vehicle ID:** Each vehicle has a unique identification number.

• **Current Road Segment:** Identifies the current road that the vehicle is using to get to the intersection.

• **Current Lane:** Identifies the lane being used.

• **Next Road Segment:** The next road taken by the vehicle after crossing the intersection.

• **Next Vertex:** The next intersection that the vehicle is getting close to.

• **Arrival-Time:** The time at which the vehicle gets to the intersection.

• **Exit-Time:** The time at which the vehicle will exit the intersection.

• **Message Sequence Number:** A unique number for each message from a vehicle. This count gets incremented for each new message generated by the same vehicle. This helps a receiver since it only needs to process the last message received from a particular sender.

• **Message Type:** The type of the message which is STOP in this case.

The CLEAR message contains 3 parameters: **Vehicle ID, Message Sequence Number**, and **Message Type:** CLEAR.

The CONFIRM message contains 3 parameters: **Vehicle ID, Message Sequence Number**, and **Message Type:** CONFIRM.

The DENY message contains 3 parameters: **Vehicle ID, Message Sequence Number**, and **Message Type:** DENY.

IMPLEMENTATION

In this section, we describe the implementation of the V2V protocols and the messages described in the previous two sections. To implement and analyze intersection protocols, traffic at intersections needs to be simulated. For this purpose, we use a tool called GrooveNet [1, 2] built at Carnegie Mellon University. We first give a brief introduction to GrooveNet and describe the extensions made.

GROOVENET[2]

GrooveNet [1, 2] is a sophisticated hybrid vehicular network simulator that enables communication among simulated vehicles, real vehicles and among real and simulated vehicles. By modeling inter-vehicular communication within a real street map-based topography, GrooveNet facilitates protocol design and also in-vehicle deployment. GrooveNet's modular architecture incorporates multiple mobility models, trip models and message broadcast models over a variety of links and physical layer communication models. It is easy to run simulations of thousands of vehicles in any US city and to add new models for networking, security, applications and vehicle interaction. GrooveNet supports multiple network interfaces, GPS and events triggered from the vehicle's onboard computer. Through simulation, message latencies and coverage under various traffic conditions can be studied.

New models can easily be added to GrooveNet without concern of conflicts with existing models as dependencies are resolved automatically. Three types of simulated nodes are supported: (i) vehicles which are capable of multi-hopping data over one or more DSRC channels, (ii) fixed infrastructure nodes and (iii) mobile gateways capable of vehicle-to-vehicle and vehicle-to-infrastructure communication. GrooveNet's map database is based on the US Census Bureau's TIGER/Line 2000+ database format [10]. Multiple message types such as GPS messages, which are broadcast periodically to inform neighbors of a vehicle's current position, are supported. On-road tests over 400 miles within GrooveNet have lent insight to market penetration required to make V2V practical in the real world [1].

Mobility Models

One major extension to GrooveNet that we made is the inclusion of lane information for roads. The TIGER map database has no information concerning the number of lanes along each road. We used the heuristic of adding lane information based on road-type information present in the database. GrooveNet has several mobility models, such as the Street Speed, Uniform Speed and Car-Following models. We have modified these models for our current purposes. In addition, we have also created new mobility models that support the presence of multiple lanes with vehicles now also having the ability to switch lanes. Cars can switch lanes either at randomly chosen times or using predefined starting lanes. Specifically, the new mobility models that were implemented are as follows:

1. Stop-Sign Model: When a simulated vehicle approaches an intersection managed by stop-signs at each entrance, it comes to a complete stop regardless of the situation of any other vehicle at the intersection. In other words, the velocity of the vehicle becomes zero even if there is no other car trying to cross the intersection. In discussions, police

---

[2] GrooveNet is an acronym that stands for "**G**eographical **R**outing **o**f **V**ehicular **Net**works".

recommend 3 seconds of complete stopping even at an empty intersection. This stop delay will increase in proportion to the number of cars that arrived earlier at the intersection.

2. Traffic-Light Model: The traffic-light model follows the same basic logic as the stop sign model except that stop signs are now replaced by traffic lights. The *Green-Light Time* of the traffic light has a default value that can be changed by the user. Both the Stop-Sign and Traffic-Light models have been designed to simulate the behavior of vehicles at intersections equipped with stop-signs or traffic-lights. In these two models, vehicles do not communicate with each other.

3. V2V Stop-Sign Model: This model represents the implementation of the Stop-Sign Protocol (SSP) described earlier. Each intersection in the map has a unique number which is called its *Vertex Number*. Based on the vertex number, each vehicle determines the next intersection it is approaching and also all the roads connecting at this intersection. The vehicle sends out a periodic safety message as described earlier. These messages are processed by other vehicles receiving them to know if multiple vehicles are approaching the same intersection. A priority-assignment policy decides which vehicle gets to cross the intersection first. In case of distinct arrival times, a first-come-first-served policy is used. In case of ties, tie-breaking rules are applied. Any vehicle with a lower priority comes to a stop at the intersection. The vehicle then checks if other vehicles have exited the intersection. Based on the V2V stop-sign protocol, if the vehicle should remain stopped, the velocity stays zero until its next update cycle, after which the tests are executed again. This continues until the vehicle gets the permission to cross the intersection and sets its velocity to the street speed limit.

4. Throughput-Enhancement Model: The Throughput-Enhancement Protocol (TEP) is implemented by this model. This model uses the complete collision detection algorithm (CDAI) including *Space Conflict*. Vehicles obey the car-following rules on the road before getting to the intersection such that their speed gets adjusted to the vehicle in front based on the information received within periodic Generic safety messages. As the vehicle arrives at the intersection, it follows the V2V-based intersection rules and uses CDAI to determine if it is safe to cross the intersection. All safety messages including STOP, CLEAR and GENERIC are sent with a frequency of 10Hz. All safety messages utilize the same 10Hz V2V Basic Safety Message (BSM) formats defined by SAE J2735 Dedicated Short Range Communications (DSRC) Message Set Dictionary. Data elements in Part II of BSM are used to specify the type of the safety message and also encapsulate related elements defined in the previous section. The safe distance maintained between two contiguous vehicles is selected to be 10 m.

5. **Throughput Enhancement with Agreement Model:** This model is designed to use all five types of safety messages: STOP, CLEAR, CONFIRM, DENY and GENERIC. Each vehicle moves based on the car-following protocol before approaching an intersection as well as after exiting the intersection area. Vehicles follow the Throughput-Enhancement with Agreement Protocol (TEPA) rules to get to an agreement on the sequence that the vehicles at the intersection should cross and also inform each other about their decision.

EVALUATION

In this section, we present a detailed evaluation of the proposed protocols using the models added to GrooveNet. The evaluation is carried out under different types of traffic scenarios and using different kinds of intersections. We compare the different mobility models: the Stop-Sign Model, the Traffic-Light model and V2V-interaction models. Two instances of the Traffic Light model are used, one with green light duration of 10 seconds and another with duration of 30 seconds.

In this paper, we do not consider any lost messages due to a lossy communication medium and we have assumed a GPS system with high accuracy. Under these assumptions, the TEP and TEP-A will behave in exactly the same manner. This also holds true for the V2V stop sign model as compared to the normal stop-sign model. Therefore, as part of our evaluation, we only consider TEP and the stop-sign model.

METRIC

We calculate the trip time for each simulated car under each model and compare that against the trip time taken by the car assuming that it stays at a constant street speed and does not stop at the intersection. The difference between these two trip times is considered to be the **trip delay** due to the intersection. We take the *average trip delays* across all cars in a simulation sequence as our metric of comparison.

The trip route for each car is calculated using the DjikstraTripModel in GrooveNet which calculates the shortest route between two points using Djikstra's algorithm. The route is chosen with a waypoint at the intersection forcing the route to pass through the intersection. The logging mechanism in GrooveNet was modified to enable logging of start and end times of cars to measure their trip times.

SCENARIOS

Since there is a large variation in intersection types, we restrict our attention to the following three categories of intersections:

• **Four-way Perfect-Cross Intersections:** The intersection legs are at perfect right angles to the neighboring leg.

• **T-junction:** Two roads are perpendicular to each other, and one of the roads ends at the intersection.

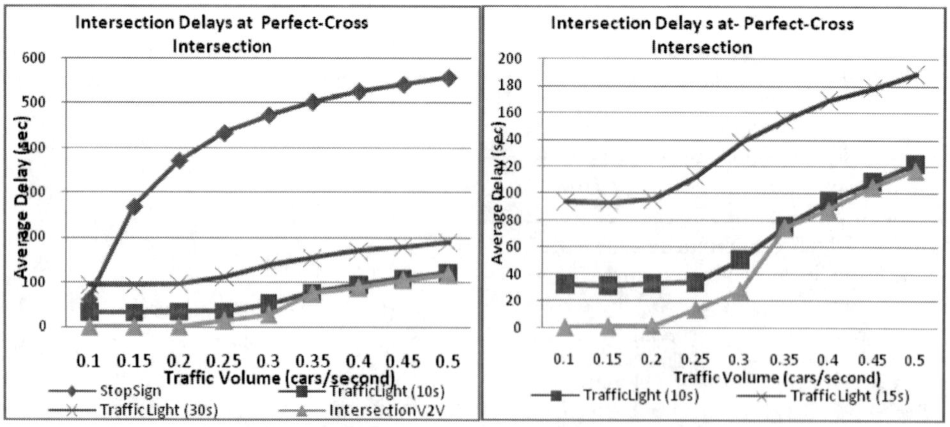

Figure 3. Delays for Perfect-Cross Intersection. Figure (a) shows all protocols. Figure (b) shows more detail w/o the Stop-Sign Protocol.

• **Four-way Angled Intersections:** These intersections are four-way intersections where they do *not* intersect at a right angle.

We run all our simulations on 4-lane roads, with 2 lanes in each direction. The intersection type, vehicle-birthing sequence, vehicle routes and turn-types are generated offline. Each vehicle is removed from simulation when it reaches its destination. This file is then fed into GrooveNet to simulate the intersection protocols. Traffic volume is specified on a per intersection-leg basis, allowing intersection legs to have different traffic levels. Each simulations uses 250 vehicles, and each run is terminated when the last vehicle reaches its destination. The simulation model in GrooveNet was modified to prevent a vehicle from becoming active if vehicles with earlier start times are already present within 10 meters of its starting position in its lane. This feature prevents cars from starting if the lane is already completely backed up.

EXPERIMENTAL RESULTS

In our first experiment, we compare different protocols for a perfect-cross intersection with an equal amount of traffic volume in every lane and an equal amount of turn ratios (that is, a vehicle has equal odds of going straight or making a turn at an intersection). The results are presented in Figure 3-(a). As can be expected, the Stop-Sign model results in higher average delays than the other protocols. As the traffic volume increases above 0.1 vehicles per second, the performance of the Stop-Sign model drops dramatically and significant traffic backlog results. In contrast, both Traffic-Light models behave at a near-constant level until the traffic volume reaches 0.25 cars per second for the Traffic-Light model with a green-light time of 10 seconds and 0.3 cars per second for the Traffic-Light model with a green-light time of 30 seconds. After that, the average delay jumps until it settles down at a higher near-constant level at about 0.35 cars per second. Beyond this traffic volume, the Traffic-Light models behave the same regardless of the green-light duration as all the lanes are completely saturated and traffic is backed up significantly. The V2V Intersection model performs the best, doing very well at low traffic volumes up to 0.2 vehicles per second resulting in very negligible delay. As traffic volume increases, the average delay increases and beyond 0.3 cars per second, it performs very similar to the Traffic-Light model with a green-light time of 10 seconds. However, the overall performance improvement is about 26% as compared to the latter Traffic-Light model. Figure 3-(b) zooms into the plot of Figure 3-(a) to show a detailed comparison between the Traffic-Light models and the Intersection model, by not showing the poorly performing Stop-Sign model.

According to classical queueing theory, the average delay will asymptotically become *very* high when the arrival rate (i.e. traffic intensity) exceeds the service rate (throughput) at the intersection. This delay, however, occurs under steady-state conditions only after a considerable amount of time. Due to practical considerations, our simulations are run for finite durations, and hence capture only transient delay behaviors after overload conditions have been reached. Nevertheless, our results clearly indicate that before overload conditions are reached, the service rate (i.e. throughput) with the V2V-Intersection protocol is noticeably better than the Traffic-Light models.

We then repeated the above experiment for a T-junction and the corresponding results are shown in Figure 4. For the T-junction, the V2V-Intersection protocol has an 83% overall performance improvement over the Traffic-Light model with a 10-second green-light time, and a 94% overall performance improvement over the Traffic-Light model with a 30-second green-light time. The T-junction has fewer conflicts to deal with than at a perfect-cross intersection, resulting in less stopping at the intersection for the V2V-Intersection model leading to its much better performance than before.

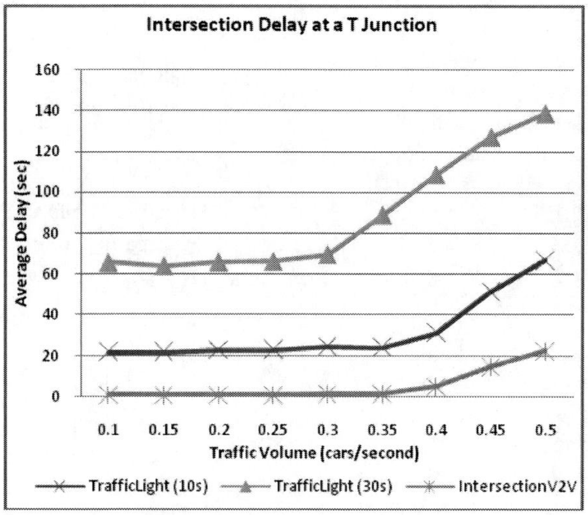

Figure 4. Delays at a T-Junction

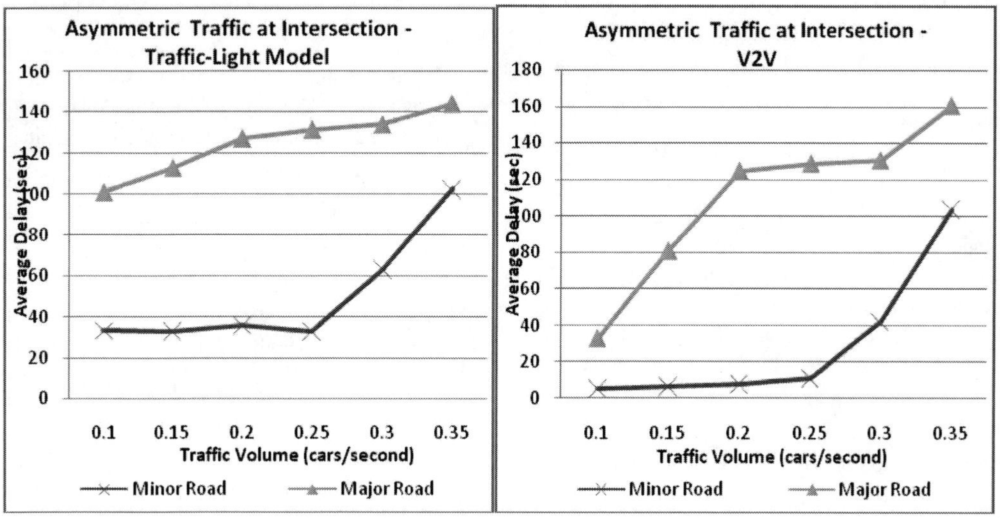

Figure 5. Delays with Asymmetric Traffic Traffic-Light Model (b) V2V Intersection Model

Next, we studied the scenario where traffic varies along different intersecting roads. That is, when two roads intersect, one road has more traffic than the other. However, we still assume that both roads have the same type and hence one does not have priority over the other. The corresponding results are given in Figure 5-(a) for the Traffic-Light Model and Figure 5-(b) for the V2V-Intersection Model. Again, the V2V-Intersection Model performs better than the Traffic-Light Model.

SUMMARY/CONCLUSIONS

A substantial fraction of automotive collisions occur at intersections. Furthermore, intersections are often traffic bottlenecks contributing to significant trip delays. In this paper, our goal was to design intersection management protocols using only vehicle-to-vehicle communications to address these two core issues of safety and throughput. We believe that intersection collisions can be reduced and throughput improved significantly using only V2V protocols. Since installing wireless infrastructure at every intersection to support vehicle to intersection protocols can be prohibitively costly, a V2V-based approach seems more practical for deployment. We have described and evaluated four V2V-based protocols namely Stop-Sign, Traffic-Light, Throughput-Enhancement and Throughput-Enhancement with Agreement protocols. We have also compared these protocols to conventional stop-signs and traffic lights, and have evaluated the average delays encountered at an intersection. We extended GrooveNet [1], a sophisticated hybrid vehicular network simulator, to support these protocols. Our results indicate the potential of these new V2V-based protocols to manage intersections with minimal dependency on infrastructure. Although our protocols are designed for autonomous vehicles that use V2V communication for co-operative driving, they can be adapted to a driver-alert system for manual vehicles at traffic intersections.

LIMITATIONS

The protocols we evaluate do not take into account any controller model for the cars. Since they assume a simplistic movement model based on current speed and current heading, certain assumptions will be violated when applied to real cars, especially when considering the throughput enhancement model where we look at polygon intersections. The ability to integrate several different controller models needs to exist and their effect on the protocols needs to be studied. Currently, we also do not deal with position inaccuracies and packet losses with wireless communication. Position accuracy will affect the protocols since each vehicle depends on its position and the known position of the other vehicles to make safety-critical decisions. Wireless packet loss results in dropped messages between vehicles and this can lead to vehicles not being able to sense other vehicles around them. We also make assumptions at a global level such as the constant speed of all cars (unless they are using the car-following model), and these assumptions are made at a global level. Hence, protocol changes will also need to be made for adapting to scenarios involving different types of cars traveling at different speeds.

Future Work

We intend to overcome the limitations described above and extend the V2V protocols in the context of real cars. We are working on hybrid simulations with real and simulated vehicles to take advantage of GrooveNet's hybrid environment. We are also working on extending our protocols to support enhancements, which will allow a vehicle to slow down, and not come to a complete stop, at an intersection to allow another vehicle to cross. We have indeed already implemented a version of this protocol on real-world Segway robots but it is not captured in this paper. There is also ongoing work to look at integration of Vehicle-to-Infrastructure (V2I) technologies within these protocols to take advantage of statically known entities at intersections. We intend to design new protocols which use the integration of V2I and V2V for managing intersections, where autonomous and human-driven vehicles are both present.

REFERENCES

1. Mangharam, R., Weller, D. S., Rajkumar, R., Mudalige, P. and Bai, Fan, "GrooveNet: A Hybrid Simulator for Vehicle-to-Vehicle Networks", Second International Workshop on Vehicle-to-Vehicle Communications (V2VCOM), San Jose, USA. July 2006

2. Mangharam, R., Meyers, J., Rajkumar, R., Stancil, D. et al., "A Multi-hop Mobile Networking Test-bed for Telematics," SAE Technical Paper 2005-01-1484, 2005, doi: 10.4271/2005-01-1484.

3. DARPA. The DARPA urban challenge, 2007.http://www.darpa.mil/grandchallenge.

4. GM, EN-V concept vehiclesat expo 2010, http://media.gm.com/content/media/us/en/news/news_detail.globalnews.html/content/Pages/news/global/en/2010/0817_env_capabilities

5. Dresner, Kurt & Stone, Peter (2008), Replacing the Stop Sign: Unmanaged Intersection Control, The Fifth Workshop on Agents in Traffic and Transportation Multiagent Systems. pp. 94-101, Estoril, Portugal.

6. Dresner, Kurt & Stone, Peter (2008), A Multiagent Approach to Autonomous Intersection Management. Journal of Artificial Intelligence Research(JAIR)

7. Drabkin, V., Friedman, R., Kliot, G., and Segal, M.. Rapid: Reliable probabilistic dissemination in wireless ad-hoc networks. In The 26th IEEE International Symposium on Reliable Distributed Systems, Beijing, China, October 2007.

8. Golshtein, E. G.; Tretyakov, N.V.; translated by Tretyakov, N.V. (1996). Modified Lagrangians and monotone maps in optimization. New York: Wiley, p. 6.

9. Shimizu, Kiyotaka; Ishizuka, Yo; Bard, Jonathan F. (1997). Nondifferentiable and two-level mathematical programming. Boston: Kluwer Academic Publishers, p. 19.

10. U.S Census Bureau. http://www.census.gov/geo/www/tiger/

11. Dresner, Kurt, Ph.D. Dissertation, "Autonomous Intersection Management", University of Texas at Austin, May 2010.

12. US Department of Transportation-Federal Highway Administration Publication, National Agenda for Intersection Safety http://safety.fhwa.dot.gov/intersection/resources/intersafagenda/

13. Meythaler, Eric, 2D Rotated Rectangle Collision http://www.gamedev.net/reference/programming/features/2dRotatedRectCollision/

14. Maile, M., Ahmed-Zaid, F., Basnyake, C., Caminiti, L., Kass, S., Losh, M., Lundberg, J., Masselink, D., McGlohon, E., Mudalige, P., Pall, C., Peredo, M., Popovic, Z, Stinnett, J., and VanSickle, S. Cooperative Intersection Collision Avoidance System Limited to Stop Sign and Trafic Signal Violations (CICAS-V) Task 10 Final Report: Integration of Subsystems, Building of Prototype Vehicles and Outfitting Intersections. Washington, DC: National Highway Traffic Safety Administration.

15. Vehicle Safety Communications 2 consortium. Cooperative Intersection Collision Avoidance System Limited to Stop Sign and Trafic Signal Violations (CICAS-V) Final Report. Washington, DC: National Highway Traffic Safety Administration.

16. Ahmed-Zaid, F., Bai, F., Bai, S., Basnayake, C., Bellur, B., Brovold, S., Brown, G., Caminiti, L., Cunningham, D., Elzein, H., Ivan, J., Jiang, D., Kenny, J., Krishnan, H., Lovell, J., Maile, M., Masselink, D., McGlohon, E., Mudalige, P., Rai, V., Stinnett, J., Tellis, L., Tirey, K., VanSickle, S. *Vehicle Safety Communications - Applications (VSC-A)* First Annual Report. Washington, DC: National Highway Traffic Safety Administration.

17. Vehicle Safety Communications 2 consortium. *Vehicle Safety Communications - Applications (VSC-A)* Final Report. Washington, DC: National Highway Traffic Safety Administration

18. Bouraoui, Laurent, Petti, Stéphane, Laouiti, Anis, Fraichard, Thierry and Parent, Michel, INRIA Rocquencourt, "Cybercar cooperation for safe intersections", Proceedings of the IEEE ITSC, 2006 IEEE Intelligent Transportation Systems Conference, Toronto, Canada, September 17-20, 2006

CONTACT INFORMATION

Reza Azimi
sazimi@andrew.cmu.edu

Gaurav Bhatia
gnb@ece.cmu.edu

Ragunathan (Raj) Rajkumar
rai@ece.cmu.edu

Priyantha Mudalige
priyantha.mudalige@gm.com

SAE International™

Integrating In-Vehicle Safety with Dedicated Short Range Communications for Intersection Collision Avoidance

2010-01-0747
Published
04/12/2010

Craig Robinson and Luca Delgrossi
Mercedes-Benz Research & Development North America, Inc.

ABSTRACT

This paper describes in detail a prototype safety system presented at the Intelligent Transportation Systems (ITS) World Congress held in New York, NY in 2008. This system relies on real-time information broadcast from the intersection and received by the vehicle over a dedicated wireless channel to detect potential red light violations. Audible and visual warnings in the vehicle are used to alert the driver. If these warnings are ignored and a red light violation is imminent, the vehicle brakes automatically and comes to a safe stop before entering the intersection.

Furthermore, this system is integrated with existing in-vehicle safety features with pre-crash and brake assist functionality. Pre-crash safety features are activated as an additional precaution, to reduce impact energy and protect the vehicle occupants in case of an unavoidable collision. Although this prototype was built for demonstration purposes only, it offers an interesting portent for the development of cooperative safety systems based on the integration of wireless communications with in-vehicle safety systems.

INTRODUCTION

Over 6 million motor vehicle crashes occur every year in the U.S., resulting in over 42,000 deaths. In 2007, an average of 112 people died each day in motor vehicle crashes (one every 13 minutes) and almost 2.5 million people were injured [1]. The annual direct economic cost to society due to injuries and damage to property is estimated to be around $230 billion. Intersection crashes, i.e., accidents occurring within the limits of an intersection or as an intersection is approached or about to be exited by a vehicle, are a large portion of fatal accidents, accounting for about 9,000 deaths every year [2]. About 250,000 accidents involve vehicles running a red light and colliding with another vehicle crossing the intersection from a lateral direction. These accidents account for $6.6 billion in economic cost [3].

Several ongoing research efforts focus on solutions to avoid or mitigate the effects of collisions associated with red light violations. One promising approach is based on active cooperation between vehicles and the roadside infrastructure. This cooperation is made possible today by the availability of wireless communication technology suitable for vehicles moving at high speeds, namely Dedicated Short Range Communications (DSRC) at 5.9 GHz [4].

Communications-based solutions have the potential to significantly enhance vehicle safety by enabling a wide range of innovative vehicle-to-infrastructure (V2I) and vehicle-to-vehicle (V2V) applications. The U.S. Department of Transportation (USDOT), building on the previous Vehicle Infrastructure Integration (VII) initiative, launched the IntelliDrive℠ program at the end of 2008 as a framework for research in this area (see www.intellidriveusa.org).

To illustrate the effectiveness of communications-based solutions and their potential benefits, Mercedes-Benz Research & Development North America, Inc. in California and Daimler Group Research in Germany integrated DSRC equipment into two research vehicles and publicly demonstrated advanced safety features at the 15th ITS World Congress. Due to the high level of integration with existing in-vehicle safety features, we refer to this system as the Integrated Safety demonstration prototype.

(2a) (2b) (2c) (2d) (2e)

Figure 2. Snapshots of the dashboard display used for demonstration purposes.

Figure 1. One of the Integrated Safety research vehicles in action at the 15th ITS World Congress in New York.

Building on a system developed as part of the Cooperative Intersection Collision Avoidance System - Violation (CICAS-V) [5] project a joint endeavor sponsored by the USDOT, the Integrated Safety demo relies on information broadcast by the intersection to determine a potential red light violation. Audible and visual warnings in the vehicle are used to alert the driver of this risk. If the system determines that the driver is not reacting to the warnings and a violation becomes imminent, onboard safety systems such as Mercedes-Benz PRE-SAFE® and BRAKE ASSIST are activated as additional measures to protect the occupants and prepare the vehicle for an imminent crash. In this demo, the vehicle comes to an automatic stop before entering the intersection. Note, however, that the system is not intended to take full control of the vehicle and the driver can override it at any time simply by pressing the brake or gas pedal.

In the following section, we provide background material on cooperative safety systems and related work on intersection collision avoidance. Then, we describe the concept of operations for the Integrated Safety demonstration and the system's main hardware and software components. A high-level overview of the technology and extensions required to implement the prototype is provided. Finally, we discuss three key implementation issues, communications, vehicle positioning, and brake torque calculation.

BACKGROUND AND RELATED WORK

Passive safety systems minimize injury of the car's occupants when a crash occurs. For instance, seat-belts prevent an occupant from being ejected and airbags inflate to reduce impact forces on the passengers. Passive safety is instrumental in the protection of passengers: more than half of the vehicle occupants killed in 2006 in the U.S. were unrestrained [2].

Active safety systems rely on in-vehicle sensors to change the response of the vehicle in an attempt to avoid a crash altogether. For example, the Electronic Stability Program (ESP) uses various in-vehicle sensors to detect a possible loss of control. Stability can then be improved by reducing engine torque and applying the brakes to one or more wheels to avoid under- or over-steering.

The introduction of wireless communication systems, such as 5.9 GHz DSRC, has the potential to further enhance active safety systems and enable cooperative applications where the vehicle interacts with the roadside infrastructure and/or other vehicles. This is particularly valuable in intersections.

The CICAS-V project is a prominent research effort towards intersection safety. CICAS-V was jointly developed by the USDOT and 5 automotive manufacturers under the Crash Avoidance Metrics Partnership (CAMP) framework [6]. In the CICAS-V application, drivers were given audible, visual, and haptic warnings when an intersection violation seemed likely. The Integrated Safety demo presented in this paper builds on and extends CICAS-V technology.

In addition to CICAS-V, the CICAS Signalized Left Turn Assistance (CICAS-SLTA) and Stop Sign Assistance (CICAS-SSA) projects have been structured to address further aspects of intersection collision avoidance.

In Europe, the AKTIV-AS Intersection Assistance project (see www.aktiv-online.org) has similar objectives: reducing the number of accidents by supporting the driver while entering, crossing, or turning into an intersection. It is expected that the AKTIV-AS solution will employ on-board sensors, communications, positioning, and digital maps.

<figure 2 here>

CONCEPT OF OPERATIONS

As the vehicle approaches, it comes within range of an interection's DSRC radio, part of what's known as Roadside

Equipment (RSE). The image in Figure 2(a) is displayed in the vehicle when no RSEs are detected and 2(b) when the vehicle receives RSE communications. The banner along the bottom of each image provides vehicle status, including brake and accelerator pedal status, activation of active safety systems, and automatic stop. Note that this driver vehicle interface (DVI) was created for demonstration purposes only.

The vehicle receives several safety messages via DSRC from the RSE. Using this information and onboard vehicle sensors (such as GPS, velocity, acceleration, and yaw rate) the vehicle can determine which intersection is being approached, what lane the vehicle is in, and the corresponding traffic signal phase and timing. If the driver is proceeding toward a red light and it appears he or she has not taken appropriate action, the vehicle warns the driver and attempts to avoid a red light violation in three stages.

RED LIGHT ADVISORY (STAGE I)

This is an advisory issued to the driver indicating that a red light is being approached and that the driver should have begun to take action already. A 'Stop Ahead' icon is displayed, as shown in Figure 2(c). A subtle acoustic warning is also issued.

RED LIGHT WARNING (STAGE II)

If the driver continues toward the intersection without taking action, the severity of the advisory is elevated, a 'Stop' icon as shown in Figure 2(d) is displayed, and a loud tone is heard. The timing of this warning is such that the driver can still react in time to stop the vehicle before entering the intersection. However, hard braking with deceleration of up to 0.6g would be required to stop in time.

AUTOMATIC STOP (STAGE III)

This is the most critical part of the warning sequence and is only reached if the driver has not responded to the previous warning stages and a red light violation is unavoidable. At this point, several actions are taken:

1. The driver is issued a loud audible warning and the vehicle display indicates that an automatic stop is being executed, as shown in Figure 2(e).

2. The vehicle prepares for a potential collision by activating the Mercedes-Benz PRE-SAFE® features. These include the tightening of seat belts, optimizing seat positions, rolling up the windows, and raising the rear headrests.

3. The vehicle applies the brakes so as to stop the vehicle before entering the intersection.

The three warning stages are adaptively applied based on the approach speed of the vehicle as well as the road conditions. They can also be adapted to the characteristics of the driver. However, it is important to note that the onset of autonomous braking is independent of driver style since it is as applied at the last possible point that the vehicle can be stopped before entering the intersection and requires no driver input. Thus, driver reaction time can be ignored in the calculation. If the driver begins to brake (or accelerate) at any time, the system is overridden. This allows the driver to maintain final control should the situation warrant.

A typical sequence of events which leads to the activation of the safety systems and finally autonomous braking is shown in Figure 3.

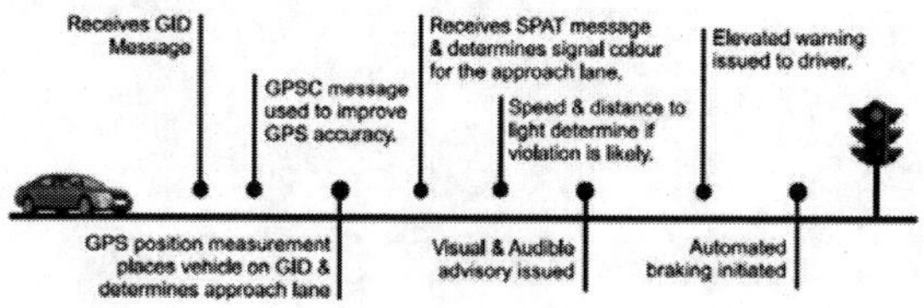

Figure 3. Typical sequence of events leading to automatic stop.

SYSTEM ARCHITECTURE

The system developed for the Integrated Safety demonstration can broadly be divided into three areas. First is the RSE, which includes the hardware and software to interface with the traffic signal controller and broadcast safety information to vehicles. The second set of components are the vehicle's active safety systems. These are production systems and are used as such in this project. The last component is the on-board equipment (OBE) responsible for receiving the broadcast from the RSE, gathering information from the vehicle network, computing vehicle position, and generating warnings. The OBE also activates the vehicle's active safety systems and implements automatic stop. These subsystems are described in detail below.

ROADSIDE EQUIPMENT

The correct behavior of the system depends fundamentally on obtaining critical information from the intersection. This is enabled by the installation of RSE interfacing with the local traffic signal controller as shown in Figure 4.

The RSE uses DSRC to broadcast safety information to approaching vehicles. The following DSRC messages are periodically broadcast by the RSE:

1). Geometric Intersection Description (GID)

This message contains a detailed map of the intersection, including number of lanes and approaches, position of stop bars, and reference points. The vehicle uses this map to position itself on an approach lane (e.g., South bound left turn lane). Also, each GID contains identification and version information so that only the most up to date maps are stored and used. At present, GIDs are created using survey equipment and stored in the RSE, but future GIDs may benefit from automated GPS tracing. GIDs are designed to be as compact as possible, thus fitting into a single DSRC message [7].

Figure 4. Roadside equipment at 36th St and 11th Ave, New York.

2). Signal, Phase, And Timing (SPaT)

This message includes dynamic traffic light status information. For each approach to the intersection, the signal color is indicated as well as the expected duration (e.g., green for 10 seconds). Knowledge about vehicle lane position is required to determine the signal color that is relevant for that position. SPAT information is represented in the form of binary XML, following the currently proposed SAE standard [7].

3). GPS Correction (GPSC)

This message contains Radio Technical Commission for Maritime Services (RTCM) differential GPS correction information which enables the on-board GPS device to obtain very accurate position estimates [8]. A stationary differential GPS (DGPS) receiver installed at the intersection acts as the base station. Using this differential scheme, lane level positioning can be achieved even with a low-cost GPS device in the vehicle.

For a typical signalized intersection, the phase timing is handled by a traffic signal controller located in a cabinet on the intersection corner. The RSE is responsible for gathering signal information from this controller, formatting the data, and broadcasting this to approaching vehicles. Thus, an RSE contains at the minimum an interface with the traffic controller (e.g., a serial or LAN connection) and a DSRC radio and antenna. The antenna is located so as to provide good coverage of the intersection and its approach lanes. When DGPS corrections are being used, the RSE also includes a DGPS base station receiver and antenna.

As using SPAT information outside of the traffic controller was never foreseen, most existing controllers do not have the functionality to provide this information in real-time. Furthermore, multiple equipment suppliers, regional traffic management authority rules, equipment age, and the policies in place during installation contribute to complicate the matter. For the Integrated Safety demonstration, we worked closely with US Traffic, a traffic signal controller supplier, as well as the New York Department of Transportation to develop an interface protocol between the RSE and the local traffic controller and install and test the equipment at intersections along 11th Ave in Manhattan.

IN-VEHICLE SAFETY SYSTEMS

For the Integrated Safety demonstration we made use of the Mercedes-Benz PRE-SAFE® system. This active safety system was first introduced in 2002 at the Paris Motor show and marked the first time that active and passive safety systems were combined into one system. It was introduced into production in the 2003 Mercedes-Benz S-Class and has since undergone a series of enhancements and extensions. The current system uses a variety of on-board sensors (e.g. radar and accelerometers) to detect hazardous conditions. When activated, Mercedes-Benz PRE-SAFE® can close the windows and sunroof (to allow the airbags to deploy optimally and prevent occupants from being ejected from the vehicle), tighten the seat belts (to remove seat-belt slack and position the occupants better for a collision), adjust the seat position and inflate seat cushions (to better position the occupants in a collision, allow better functioning of the seat belt, and prevent 'submarining' in which occupants slide off the front of the seat), and raise the rear headrests (to prevent whiplash). When a collision is inevitable, the Mercedes-Benz PRE-SAFE® brake system can initiate partial braking (1.6 seconds before impact) and full braking (0.6 seconds before impact) in order to reduce collision impact energy.

ON-BOARD EQUIPMENT

One of the objectives of the Integrated Safety demonstration is to illustrate how communications-based safety applications can be cleanly integrated into a typical driving experience. Consequently, the internal and external appearance of the research vehicles was minimally modified. However, to implement all of the system functionality, additional hardware components were installed on the vehicles (summarized in Figure 6). Of particular interest are the following system components:

Computing Platform

The DENSO Wireless Safety Unit (WSU) was used as the system computing platform. The WSU processor is a 400 MHz MPC5200B PowerPC with 128MB DDR SDRAM and runs Linux. The WSU was developed especially for automotive ITS applications and has hardware and software interfaces for two DSRC radios, GPS, and two CAN busses. To avoid overloading the platform, a second WSU was used to generate the rich audio and video Driver Vehicle Interface (DVI) for the demo.

DSRC Antenna

A DSRC antenna was integrated into the existing shark fin radome on the roof of the vehicles. The design accounted for the glass roof of the S-Class as well as interference with the GPS and GSM antennae located within the radome. Simulations on antenna signal reception characteristics were performed and a typical output is shown in Figure 5.

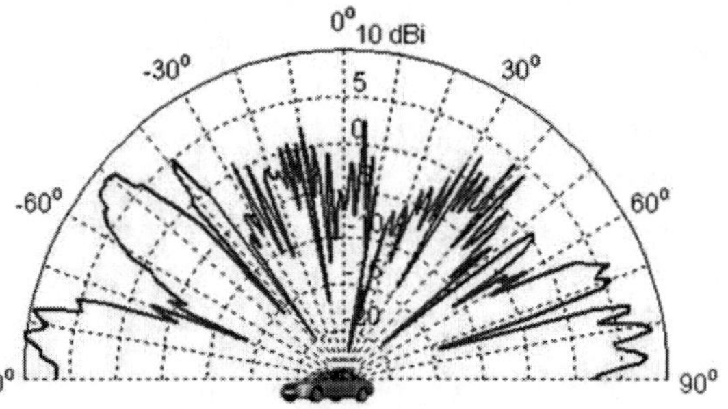

Figure 5. Signal reception strength for the DSRC antenna (simulation study).

GPS Antenna

Different GPS antenna arrangements were used in the two Integrated Safety vehicles. In the S350, the production roof mounted antenna was used. In the S550, a low cost L1 antenna was mounted on the trunk. Both antennae were connected to differential L1 GPS receivers in the trunk.

ESP ECU

The vehicle Electronic Stability Program (ESP®) Electronic Control Unit (ECU) can initiate braking with a limited maximum deceleration. For demonstration purposes, an ESP unit with specialized software was installed which allowed much larger deceleration. This mode of operation was only active during the demonstration and was not permitted for general use on public roads.

CAN Interface Gateways

Two separate system functionalities were required for the demo. The first is the activation of PRE-SAFE®, the second is autonomous braking. In order to implement the Mercedes-Benz PRE-SAFE® activation, CAN messages from the ESP ECU were intercepted before reaching the vehicle CAN bus. When required, messages were modified before being passed onto the CAN network. In order to activate autonomous braking, a similar gateway was created between the Radar Decision Unit (RDU) and the vehicle CAN. When autonomous braking was required, a CAN message was created which requested the desired braking torque (and engine braking torque) to be implemented by the ESP unit.

Emergency Shutoff Switch

In case of any system failure, an emergency power shutoff was installed into the dashboard. This switch cuts power to the ESP ECU, the WSUs, as well as the CAN interface devices so as to immediately disable the demo system while still permitting vehicle operation.

Off-board Video and DVI Display

To provide roadside spectators at the World Congress with a view of the in-vehicle display and of PRE-SAFE® activation, a wireless system was developed to stream video to a large video screen on the roadside. The video showed seat belt tensioning and seat position adjustment. A real time view of the onboard DVI was also displayed on the roadside video screen. This was achieved by transmitting a very short (2 byte) vehicle status message to a client application on the roadside which used it to generate an exact replica of the in-vehicle DVI.

On-board Software Systems

The on-board software can be broken into three separate categories. First, the software which interfaces with the vehicle and sensors was implemented on the CAN interface devices. Second, the warning algorithm logic was based on the platform developed by the CICAS-V project [6] and used many of the application interfaces provided by the WSU hardware and software drivers. Finally, the DVI software developed for both on-board and off-board applications was written in a combination of Java, C and Qt.

One important design decision was to completely separate (both physically and logically) autonomous braking from the warning generation. In this way, a low-level safety check could be performed before braking was initiated, and a well defined recovery mechanism was put in place to deal with any higher level system malfunction. In brief, red light violations were computed in one program and autonomous braking was implemented in another. The two processes communicated through only a CAN interface and failure of one system did not cause failure of the other.

SUBSYSTEMS AND ISSUES ADDRESSED

COMMUNICATIONS

Data exchange between the intersection and vehicles take place via DSRC [4]. In the U.S., DSRC operates at 5.9 GHz over a 75 MHz spectrum allocated by the U.S. Federal Communications Commission (FCC) [9]. The frequency

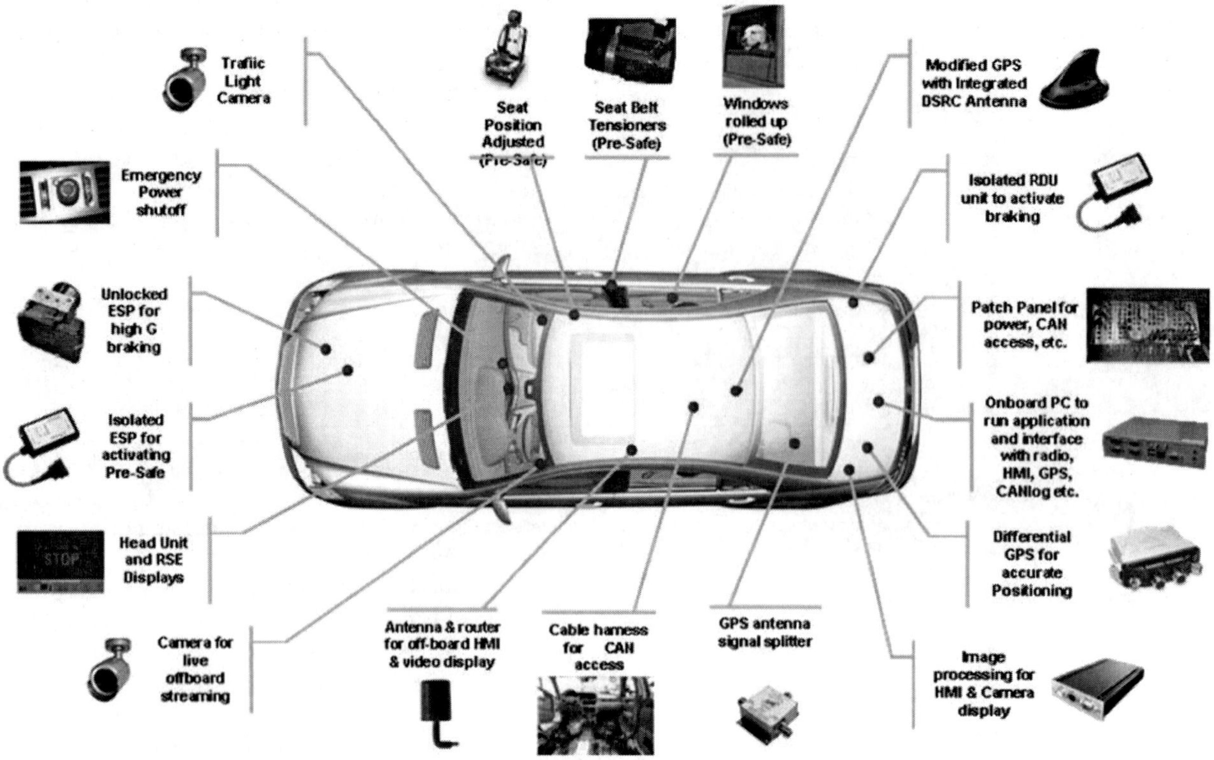

Figure 6. Illustration of the overall system hardware

allocation in Europe is similar, although only a narrower 30 MHz band is available [10].

In the U.S., the DSRC band is structured into a 10 MHz control channel (CCH) and several service channels (SCH). The CCH is primarily used for safety. It is also used to advertise available services on service channels so that radios can switch channels if necessary. Switching to a service channel is allowed for a limited time, but radios must frequently tune back to the CCH so that safety information is not missed.

The DSRC messages used for the Integrated Safety demonstration include GID, SPAT, and GPSC. GID and GPSC are not strictly safety messages and could be broadcast on a service channel to reduce control channel load. SPAT, on the contrary, includes safety information generated in real-time and should be broadcast over the CCH. Although the vehicles were equipped with DSRC radios capable of channel switching, in the New York event all DSRC messages were broadcast over the CCH.

<figure 6 here>

DSRC performs best under line of sight (LOS) conditions. The portion of 11th Ave between 34th and 38th Street is free from obstructing trees or other obstacles and vehicles had no problem receiving messages from multiple RSEs. With careful choice of locations for the roadside antennae we typically experience around 500m communication range.

POSITIONING

The Integrated Safety demonstration requires accurate lane level vehicle positioning to determine the correct approach lane and relevant traffic signal. Also, the distance to the stop bar must be accurately computed to determine when warnings should be issued and most importantly, when to initiate automated braking.

A stationary DGPS receiver is mounted at the roadside and computes satellite ranging errors. These errors are communicated via the GPSC message to nearby vehicles. Standalone GPS accuracy is presented in Table 1. We assumed worst case for each error range and a baseline of 1km. These values exclude the effect of Dilution of Precision which typically increases the expected error. From these values it is clear that using differential GPS corrections would provide sufficient positioning accuracy to support violation warnings. However, what is not fully reflected in the table are the effects of signal multi-path and interference which cause brief, yet large position errors, as shown in Figure 7.

To overcome these errors, a Kalman Filter position estimator was constructed. The estimator efficiently combines the GPS position information (latitude, longitude, velocity, Horizontal Dilution Of Precision (HDOP), and standard deviation of error in latitude and longitude) with sensor measurements available in the vehicle (wheel rotation speed and direction, yaw rate, and longitudinal acceleration) to produce an accurate vehicle position estimate. Figure 7 shows two 'walls' with each panel representing a GPS measurement. The taller wall is the raw GPS measurements obtained from the receiver, and the output from the estimator is shown as the shorter wall. The vehicle was modeled as a free body object, with lateral and longitudinal model error covariance chosen to reflect vehicle non-holonomic behavior. Model error covariance was chosen to optimize the estimator performance on a set of observed driving data. Using this model and the on-board sensor information, open loop position prediction produced errors of less than 2 meters over 200m of driving in regular driving conditions (i.e., smooth asphalt roads, moderate speeds, and low wheel slippage).

Table 1. 1σ pseudo-range error budget for GPS with and without DGPS correction.

Error Source	L1 GPS	L1-L2 GPS	L1-L2 DGPS
Broadcast clock error	1.1	1.1	0.0
L1 P(Y) - L1 C/A Group delay	0.3	n/a	n/a
Broadcast ephemeris	0.8	0.8	0.6mm
Ionospheric delay	7.0	0.1	4cm
Tropospheric delay	0.2	0.2	4cm
Receiver noise & resolution	0.1	0.1	0.1
Multipath	0.2	0.2	0.3
Total Error (RSS)	**7.1m**	**1.4m**	**0.3m**

Position updates are received from the GPS receiver at 10 Hz. Vehicle sensor updates are received at 50 Hz. None of the measurements are synchronized. Hence, a GPS position update, wheel velocity measurement, and vehicle yaw rate measurement may arrive at the estimator in a random order with different inter-arrival timing. To account for this, whenever any measurement is received, the Kalman filter time update is computed for the time period since the last measurement arrival. A measurement update is then performed using only the received measurement.

The estimator resolved two important issues. First, when insufficient satellites are visible to form a position estimate, or if the position error is very large, the GPS receiver does not provide a position estimate at all. The estimator filled in these missing observations by generating a position estimate using the on-board vehicle sensors and the vehicle model. Under this approach, different measurement updates were performed depending on which sensor measurements were received. Thus, when GPS position updates were not received, the vehicle velocity sensor updates ensured that the vehicle velocity estimation error would remain bounded even though the position error would grow unbounded.

Figure 7. Illustration of GPS Errors due to nearby object interference with satellite ranging.

Second, when an individual satellite signal is corrupted (e.g., due to interference) the HDOP and measurement error covariance values do not always reflect this condition immediately; there is sometimes a delay in determining that GPS positions reported by the receiver are inaccurate. The estimator mitigates the effect of these errors by independently evaluating the reliability of a measurement before using it to generate an updated position estimate.

BRAKE TORQUE CALCULATION

The Mercedes-Benz S-Class has two mechanisms for applying brake torque. The 'brake pump' is used by the Electronic Stability Program (ESP) unit to implement features such as Hill Start Assist, Fading Brake Support, and Airgap reduction. The 'brake booster' is used by the Brake Assist System (BAS) to augment the driver's applied brake force in emergency braking situations. For the Integrated Safety demonstration the brake pump was used to generate braking torque.

Typical brake pump behavior is illustrated in Figure 8. A brake torque request is sent via the CAN bus to the ESP unit, which activates the brake pump. This incurs a delay of approximately 0.15s during which no brake torque is applied. The pump then begins to build up pressure and the applied brake torque increases linearly. Once the desired torque has been achieved the pump maintains the required pressure until the torque request is completed.

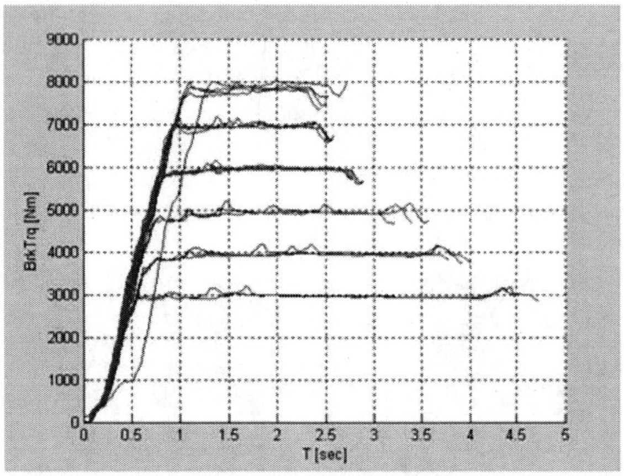

Figure 8. Illustration of typical brake torque characteristics. A brake torque request is made at time t = 0. There is a delay before the brake pump begins to apply pressure. The brake torque then increases linearly until the required torque is achieved at which point it is held constant.

It is significant to note that the brake pump attempts to deliver a requested hydraulic pressure to the brake system, which then results in a brake torque being applied. However, if the torque is too large and the wheel begins to slip, the ESP detects the slip and will reduce the torque applied to that particular wheel; the ESP remains active throughout the demonstration. Consequently, predicting a stopping distance and the associated required braking torque is very difficult if the wheel slips. For this reason, the automatic stopping distances were chosen so as to avoid excessive wheel slip. Braking distances were chosen for wet and dry conditions and a good quality asphalt surface was assumed for the demo.

Further testing revealed the relationship between brake torque and deceleration for the vehicles used in the demo. The values were different due to factors such as different engine size and total vehicle mass. This torque deceleration relationship was combined with the brake pump model above to construct a braking distance and brake torque model. The relationship is shown in Figure 9.

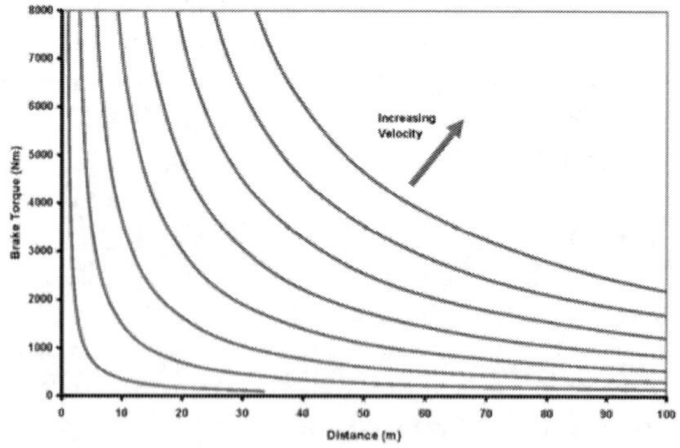

Figure 9. Illustration of the relationship between brake torque, stopping distance and velocity.

The strictly monotonic convex nature of the model lends itself well toward gradient based solution methods. A numerical Newton-Raphson gradient descent algorithm was implemented to find the braking torque required for the specified braking distance.

The algorithm and model described above formed the basis of the model predictive controller used to stop the vehicle at the stop bar. Control input was based on two different position estimates computed on different time scales by different estimators. The 'outer loop' estimator (slower time scale) represents the position estimator discussed in an earlier section. This estimator provides a new position estimate every 100ms from which a distance to the stop bar is computed. This updated value is sent via the CAN bus to the second estimator called the 'inner loop' estimator.

This estimator is tightly coupled with the brake torque control loop and maintains a distance to stop bar estimate which is updated every 20ms using wheel rotation information from the CAN bus. When an updated estimate is received from the outer loop position estimator, the inner loop estimate is replaced with the new position estimator value. The two estimators are not synchronized and intervals between outer loop estimator arrivals and inner loop prediction time points are spanned using a zero order hold on vehicle acceleration.

Using the inner loop distance to stop bar calculation, a new required brake torque is calculated every 20ms. However, due to limitations on the update frequency of the brake pump, a new brake torque could only be requested once a steady state brake torque had been achieved by the brake pump. Subsequently, a braking event from 50km/h would typically have only 5 brake torque updates.

In general, we found that the initial torque request was too little and was followed by a far larger second requested torque. Subsequent requests returned to the initial level. This indicates that further system modeling could be done to improve performance. However, the scheme reliably stopped the vehicle within 20cm of the stop bar from any initial speed.

CONCLUSIONS

The system described in this paper was demonstrated at the 15th ITS World Congress held in New York. For the duration of the event, the portion of 11th Ave between 34th and 38th Street was blocked for several hours each day and dedicated to demonstrations. Each day, a 30 minute slot was dedicated to the Integrated Safety demo. We were able to show the system operating on real roads and interacting with real traffic controllers to over 100 passengers. The demo effectively highlighted the potential for cooperative systems to improve safety, especially in the presence of an intelligent roadside infrastructure.

The ITS World Congress was broadly covered by the media and several videos and interviews were published online with images taken from the vehicles during the demonstrations. The integration of in-vehicle safety systems with DSRC will continue to be a prominent research topic. The vehicle will receive data from several sources (onboard sensors, CAN, DSRC, and GPS) and this data will be processed and utilized to further enhance the driving experience and increase safety. The information distributed by the roadside may be used for other purposes as well, including enhanced fuel economy and increased mobility.

ACKNOWLEDGMENTS

The authors would like to acknowledge the invaluable support provided by colleagues at Mercedes-Benz Research & Development North America and Daimler Group Research in Boeblingen and Sindelfingen, Germany. This demonstration would not have been possible without the valuable insights and experiences of colleagues in the Vehicle-Centric Communications team and the various active safety systems research groups within Daimler and Mercedes-Benz. Lastly, we appreciate the opportunity to build on the efforts of the USDOT and the joint CICAS-V team in order to demonstrate a state-of-the-art safety system while maintaining cross-platform compatibility.

REFERENCES

1. NHTSA National Center for Statistics and Analysis: "Traffic safety facts 2007 data", Tech. Rep. DOT HS 810 993, National Highways Transportation Safety Authority (NHTSA), 2008.

2. National Center for Statistics and Analysis (NCSA): "Traffic safety facts 2006: A compilation of motor vehicle crash data from the fatality analysis reporting system and the general estimates system", Tech. Rep. DOT HS 810 818, National Highways Transportation Safety Authority (NHTSA), 2006.

3. Najm W., Smith J., and Yanagisawa M.: "Precrash scenario typology for crash avoidance research", Tech. Rep. DOT HS 810 767, National Highway Traffic Safety Administration (NHTSA), Washington, D.C., 2007.

4. Gallagher, B., Akatsuka, H., and Suzuki, H., "Wireless Communications for Vehicle Safety: Radio Link Performance and Wireless Connectivity Methods," SAE Technical Paper 2006-21-0030, 2006.

5. Maile M. and Delgrossi L.: "Cooperative Intersection Collision Avoidance System for Violations (CICASV) for avoidance of violation based intersection crashes", 21st International Conference on the Enhanced Safety of Vehicles (ESV), no. 09-0118, June 2009.

6. Maile M., Ahmed-Zaid F., Basnyake C., Caminiti L., Kass S., Losh M., Lundberg J., Masselink D., Mc-Glohon E., Mudalige P., Pall C., Peredo M., Popovic Z., Stinnett J., and VanSickle S.: "Final Report: Cooperative Intersection Collision Avoidance System for Violations (CICAS-V)", Tech. Rep. (in print), National Highway Traffic Safety Administration, Washington, D.C., 2008.

7. SAE International Surface Vehicle Standard, "Dedicated Short Range Communications (DSRC) Message Set Dictionary," SAE Standard J2735, Iss. Dec. 2006.

8. "Differential Global Navigation Satellite System (GNSS) services," Tech. Rep. RTCM Std. 10403.1, Radio Technical Commission for Maritime Services (RTCM), October 2006.

9. Jiang D., Taliwal V., Meier A., Holfelder W., and Herrtwich R.: "Design of 5.9 ghz DSRC-based vehicular safety communications", IEEE Wireless Communications magazine, vol. 13, October 2006.

10. Festag A. and Hess S.: "News from European standardization for intelligent transportation systems", Global Communications Newsletter, June 2009.

11. Kaplan E. D. and Hegarty C. J.: "Understanding GPS Principles and Applications". Norwood, MA, Artech House, 2 ed., 2006.

CONTACT INFORMATION

craig.robinson@daimler.com

luca.delgrossi@daimler.com

The Engineering Meetings Board has approved this paper for publication. It has successfully completed SAE's peer review process under the supervision of the session organizer. This process requires a minimum of three (3) reviews by industry experts.

ISSN 0148-7191

doi:10.4271/2010-01-0747

Positions and opinions advanced in this paper are those of the author(s) and not necessarily those of SAE. The author is solely responsible for the content of the paper.

SAE Customer Service:
Tel: 877-606-7323 (inside USA and Canada)
Tel: 724-776-4970 (outside USA)
Fax: 724-776-0790
Email: CustomerService@sae.org
SAE Web Address: http://www.sae.org
Printed in USA

SAE International

2009-01-0168

Intelligent Vehicle Technologies that Improve Safety, Congestion, and Efficiency: Overview and Public Policy Role

Eric C. Sauck
University of Michigan

ABSTRACT

At the forefront of intelligent vehicle technologies are vehicle-to-vehicle communication (V2V) and vehicle-infrastructure integration (VII). Their capabilities can be added to currently-available systems, such as adaptive cruise control (ACC), to drastically decrease the number and severity of collisions, to ease traffic flow, and to consequently improve fuel efficiency and environmental friendliness. There has been extensive government, industry, and academic involvement in developing these technologies. This paper explores the capabilities and challenges of vehicle-based technology and examines ways that policymakers can foster implementation at the federal, state, and local levels.

INTRODUCTION

MOTIVATION - The American driving public has rapidly increased road use over the past decade. According to the Federal Highway Administration, between 1996 and 2006, the number of vehicle miles traveled increased by 21%, from 2,497,901 million to 3,033,753 million. Several conditions can be partly attributed to this growth: the number of traffic fatalities has remained constant, despite better safety technologies, congestion has increased, and the environmental impact of road transportation has increased. These issues translate into a significant economic loss.[1]

Safety - Despite the 17% *decrease* in traffic fatality rate—measured in deaths per 100 million vehicle miles traveled (VMT)—between 1996 and 2006, the total number of traffic fatalities has remained relatively constant at approximately 42,000 deaths per year (see Figure 1). To put this number in perspective, it is equal to one fully-loaded 747 jetliner crashing every four days. In 2004, motor vehicle collision was the number-one cause of death for Americans between the ages of 2 and 34.[2,3]

The Department of Transportation's National Highway Transportation Safety Administration (NHTSA) estimates the economic cost of motor vehicle crashes in 2000 to be $230.6 billion, including property, medical, productivity, and other losses.[3]

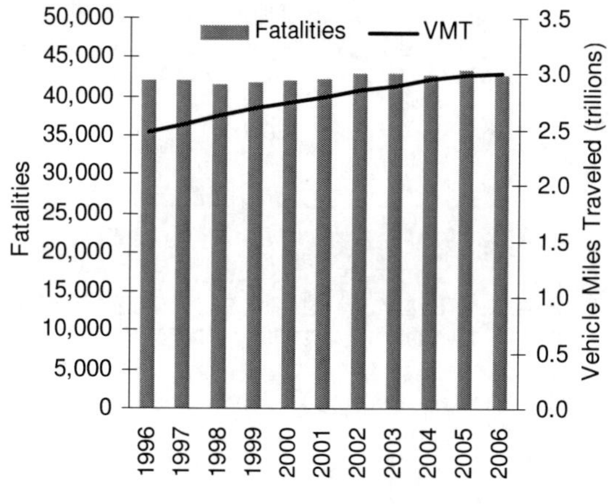

Fig. 1: U.S. motor vehicle related fatalities and vehicle miles traveled from 1996 to 2006[1]

Congestion - While the total number of vehicle miles traveled increased 21% between 1996 and 2006, the number of roadway lane-miles barely grew by only 3%. Furthermore, the U.S. Department of Transportation (DOT) projects that the use of combined road and rail will

The Engineering Meetings Board has approved this paper for publication. It has successfully completed SAE's peer review process under the supervision of the session organizer. This process requires a minimum of three (3) reviews by industry experts.

ISSN 0148-7191

SAE Customer Service: Tel: 877-606-7323 (inside USA and Canada)
Tel: 724-776-4970 (outside USA)
Fax: 724-776-0790
Email: CustomerService@sae.org
SAE Web Address: http://www.sae.org

9-2009-01-0168

Printed in USA

SAE International

increase by 250% by 2050, while roadway lane-miles will increase by only 10%. Americans are experiencing the effects of this every day in traffic congestion and delay, which has risen since 1982, according to studies by the Texas Transportation Institute (TTI) (see Figure 2). The difference will have to be resolved by advanced technology, transit, and operations management.[1,4]

The TTI 2007 Urban Mobility Report estimates that, in 2005, congestion in U.S. cities has caused people to lose 4.2 billion hours of their time and to waste 2.9 billion gallons of fuel, equating to an economic loss of $78 billion.[4]

Fuel efficiency and environmental friendliness - When traffic is congested, fuel is wasted. This unnecessary fuel burn has a direct correlation to unnecessary greenhouse gas (GHG) emissions in congestion, which have been rising from 1982 to 2005 (See Figure 2).

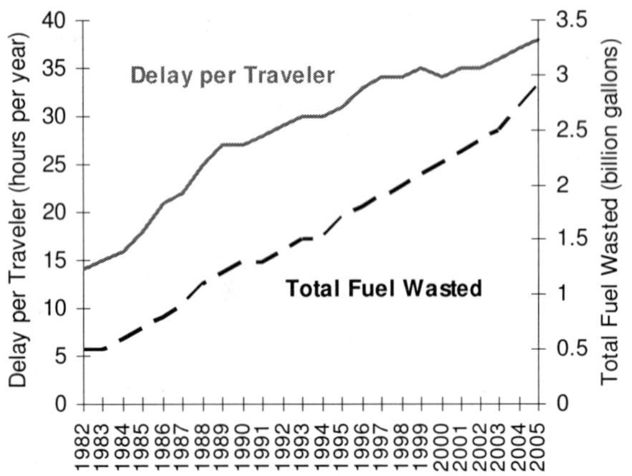

Fig. 2: Annual delay per traveler and annual total fuel wasted from 1982 to 2005[4]

In all light-duty travel, the U.S. Environmental Protection Agency (EPA) estimates that GHG emissions have gone up by 19% from 1990 to 2003. The EPA attributes this growth to both an increase in vehicle miles traveled and a limited improvement in fuel economy associated with an increase in the proportion of light-duty trucks to passenger cars. During the same period, GHG emissions from heavy-duty vehicles, including freight-carrying trucks, increased by 57%.[5]

The economic cost of GHG emissions to society is difficult to quantify. The consumption of foreign oil has national security and market collusion implications. Increased greenhouse gases have become accepted as a man-made cause of global warming with longer term impacts on society, including increased health problems.

SOCIETAL DEMANDS - In recent years, there has been a popular demand in the U.S. to decrease GHG emissions from automobiles and to decrease transportation costs due to high fuel prices. Consequently, government-mandated Corporate Average Fuel Economy (CAFE) standards were increased through the Energy Independence and Security Act of 2007. The new rules require each manufacturer's fleet of light-duty automobiles to attain a combined average of 35 miles per gallon by 2020. These new CAFE standards will require new ways to approach fuel economy; traditional methods like improving drivetrain efficiency will likely not be enough to feasibly satisfy CAFE requirements.[6]

Improvements in drivetrain efficiency have traditionally been counteracted by, among other things, additional safety features, which usually add to vehicle mass. Moreover, increases in EPA-rated fuel economy can be nullified by congestion, since vehicles should ideally travel between 40 and 60 mph (and at constant speed) to achieve optimal fuel economy. More massive vehicles, along with more frequent stop-and-go traffic cycles, are especially detrimental to fuel economy, and therefore increase GHG emissions.[7]

Because of these trade-offs, it is especially desirable and timely to implement technology that addresses all three, sometimes opposing, issues: safety, congestion, and efficiency.

SCOPE - Although there have been many developments in transportation infrastructure technology—such as signal coordination, roadway surfacing, and intersection design—this paper will focus on vehicle-based technology and its interface with infrastructure.

VISION ZERO - The goal of vehicle safety is to adopt systems that prevent collisions from occurring altogether. This so-called "Vision Zero" will require advanced *active* safety devices that entail significant technical, political, organizational, and societal challenges.

The achievement of Vision Zero will have many long-term implications for our concept of a vehicle. When vehicles no longer crash, and when anti-crash systems are proven to be fail-proof, automakers will be able to remove passive safety devices—devices such as air bags, crumple zones, and eventually, seat belts—thereby saving mass, fuel economy, affordability, and complexity.

ROADMAP: CURRENT, SHORT-, AND LONG- TERM INTELLIGENT VEHICLE TECHNOLOGY

The automotive industry has been developing safety technology at a rapid pace. For active safety systems that also mitigate congestion, the potential capabilities, benefits, and associated challenges will be briefly discussed.

CURRENT TECHNOLOGY: DRIVER WARNING AND ASSISTANCE - The market currently offers an array of driver warning and driver assistance aids to consumers.

Driver warning aids are meant to alert the driver in unsafe situations. Driver assistance aids help the driver to perform driving tasks more safely.

Current driver warning aids include the following:

- *Forward collision warning (FCW)* uses forward-facing laser or radio waves to identify an imminent crash; depending on the system, it can warn the driver, pre-load the brakes, close the windows, and tighten the seat-belts. In some cases, it can apply the brakes, as well.
- *Lane departure warning (LDW)* uses a forward-facing camera to identify lane markings and warn the driver using visual, audio, and haptic (touch) feedback.
- *Blind-spot warning (BSW)* uses cameras and radar to recognize vehicles in the driver's blind spot. If the driver begins to merge when a vehicle is in the way, the system will warn the driver via a warning light or chime.

Current driver assistance aids include the following:

- *Adaptive cruise control (ACC)* uses laser or radio waves to determine the distance, speed, and acceleration difference between the subject vehicle and a vehicle preceding it to keep a safe following distance while maintaining a preset speed whenever possible. It automatically applies throttle and braking as necessary. Some systems are capable of functioning at all speeds, while others work only above a minimum speed.
- *Lane-keep assist (LKA)* usually employs a forward-facing camera to identify lane markings, in conjunction with active steering or brake assist to maintain the vehicle in its lane.
- *Self-park system* uses cameras and short-range ultrasonic sensors to identify a parking space and automatically guide the vehicle into it by controlling the steering, acceleration, and braking.

The intended benefits of these systems are to increase driver comfort, convenience, and safety. However, ACC also acts to diminish congestion.

Safety - It is clear that an ACC-equipped vehicle can react many times quicker than a human driver, and that the control system will not overreact like humans in changing traffic conditions. These advantages may prove to have a multiplier safety effect; that is, smoother driving of the ACC-equipped vehicle will make driving in surrounding vehicles safer.

Safety benefits of driver aids in the context of NHTSA-reported related factors in fatal accidents are highlighted in Table 1.

Congestion - According to simulations done by California Partners for Advanced Transit and Highways (PATH), ACC can moderately increase single-lane roadway capacity from the current 2050 vehicles per hour with manual control to 2200 vehicles per hour with ACC. It is important to consider that roadway capacity is affected by many variables and that these numbers are intended only for comparison purposes.[8]

Challenges - With the introduction of these driving aids, there is a danger that the driver will lose attentiveness or gain a false sense of trust in the vehicle in the event of an accident. Also of concern is whether the driver will react predictably to the warning systems. Human factors concerns will be discussed further in following pages.

Another challenge is that these features are currently available mainly on premium vehicles; historically, safety features have "trickled down" to more mainstream vehicles when the cost of the technology has decreased. However, even on premium vehicles, safety features are often bundled with luxury and convenience items as option packages for marketing and cost reasons.

SHORT-TERM TECHNOLOGY: ADDING VEHICLE-TO-VEHICLE COMMUNICATION FUNCTIONALITY - Vehicles will soon be able to communicate with each other using a system called Dedicated Short-Range Communications (DSRC). The Institute of Electrical and Electronics Engineers (IEEE) is finalizing a standard, 802.11p, for wireless inter-vehicle communication. It is similar to wireless computer networking, and it uses a 5.9 GHz frequency band allocated by the Federal Communications Commission.[9]

A vehicle equipped with DSRC is capable of sharing information—position, velocity, acceleration, and other data, like braking capability—with other nearby vehicles over the secure, "ad-hoc" network. When integrated with ACC, the system becomes *cooperative* adaptive cruise control (CACC). Some of the potential improvements in safety, congestion, and fuel efficiency and environmental friendliness are described below.

Safety - The introduction of vehicle-to-vehicle communication (V2V) using DSRC could potentially eliminate certain types of crashes when combined with driver aids. It can mitigate eight of the top ten related factors in traffic fatalities identified by NHTSA (see Table 1).

Congestion - Full market penetration of CACC can further improve single-lane capacity over ACC alone to 4550 vehicles per hour, according to PATH simulations. This capacity is highly dependent on the preset following distance between vehicles, which can be increased or decreased based on industry consensus and/or on driver comfort. Figure 3 shows the simulation results for various market mixes of manual control, ACC, and CACC. Note that high market penetration of CACC is necessary to achieve significant increases in roadway capacity.[8]

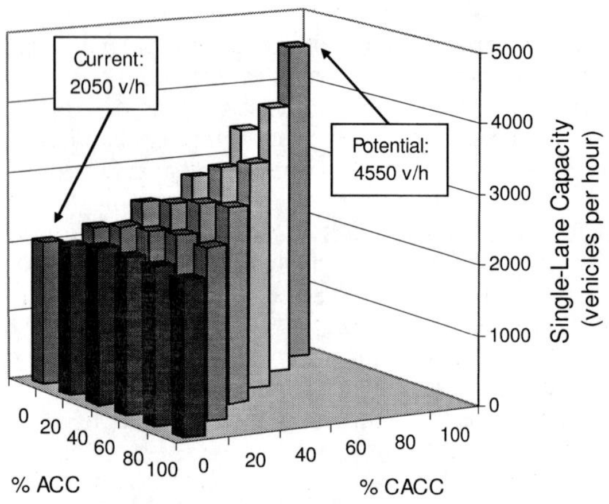

Fig. 3: Effect of ACC and CACC mix on lane capacity (reproduced with permission from author)[13]

In another model, similar benefits were found: a traffic simulation by the Netherlands-based TNO research institute found a more than 15% decrease in delays when one-half of the vehicles were equipped with CACC.[11]

Fuel efficiency and environment - While CACC can "smooth out" velocity changes in traffic and therefore improve fuel economy by maintaining more consistent speeds, the fuel efficiency benefits of V2V extend to steady-speed cruise. In a "platoon," a closely-spaced string of vehicles, aerodynamic drag force is reduced. Using CACC, the vehicle spacing can be safely shortened enough to improve fuel efficiency at highway speeds.

The effect is especially significant in freight trucking. According to the Department of Energy's 21st Century Truck Partnership, aerodynamic drag consists of 53% of the non-engine energy losses of a heavy truck at 65 mph.[12]

Challenges - Although V2V affords clear benefits to congestion, safety, fuel efficiency, and the environment, the added benefits would be limited at low market penetration of the technology, as shown in Figure 3. One possibility to more quickly attain the advantages of V2V is to retrofit all existing vehicles with DSRC transceivers, which would allow them to send their position, velocity, and acceleration status to vehicles with driver aids. When retrofit vehicles broadcast their actions, other vehicles with driver aids can help their drivers to avoid colliding with the retrofitted vehicle.

Technical challenges remain in creating efficient algorithms and processing power to handle the vast amounts of transmitted and received data. Field trials are also required to more accurately predict the effect of CACC systems in mixed (DSRC- and non-DSRC-equipped) traffic.

LONG-TERM TECHNOLOGY: ADDING VEHICLE-INFRASTRUCTURE INTEGRATION FUNCTIONALITY - Beyond V2V with CACC, the next stage in intelligent vehicle technology is vehicle-infrastructure integration (VII), where strategically-placed roadside equipment with DSRC sends data to and receives data from vehicles. It can also send it to third parties, like a traffic management office for usage statistics or a communications provider for car-based Internet access.

Safety - There are many potential safety benefits to equipping roadways, intersections, and signals with DSRC capability. In addition to extending the functionality of CACC-type systems, VII can allow drivers to receive in-car (or out-of-car) warnings of impending traffic signal violations, curve speed warnings, notices of upcoming traffic congestion and re-routing guidance, and weather alerts.

Congestion - Sophisticated road management functions can be automatically controlled with VII. Controllers can optimize highway on-ramp metering and signal priority for maximum traffic throughput based on real-time conditions.

Furthermore, VII can allow road managers to collect traffic flow information. If traffic flow indicates a problem, managers can pinpoint the problem area and quickly deploy emergency response, road maintenance, or snow removal crews.

Another aspect of VII that can potentially improve congestion is open-road tolling. Without having to stop at a toll both (a system still in use in many states), drivers will pay precisely for the road they use.

Fuel efficiency and environment - One way to improve fuel efficiency with added VII capability is to integrate three-dimensional, or topological, road maps, precise GPS location, and ACC. In a 2006 study by researchers at Linköping University in Sweden, heavy trucks were equipped with a sort of road-predictive cruise control. Based on road slope, an on-board computer selected speeds and transmission gears for optimal fuel efficiency. In their simulation along a 127-km stretch of Swedish highway, trucks reduced fuel consumption by 2.5% without adding to travel time.[13]

Topographic maps already exist for the U.S., through Geographical Information System (GIS). Road maps can be overlaid onto the topological maps. If corrections need to be made to the overlay, then vehicles equipped with GPS and DSRC can potentially upload the new road geometry values to a database. This database can be distributed to vehicles via VII and used by their ACC systems to further improve fuel efficiency.[14]

Challenges - Funding will be a major hurdle in implementing VII. Federal Highway Administration researchers estimate the cost to equip all intersections around the country with DSRC to be on the order of billions of dollars, not including expansion to other parts of roadways, like dangerous curves or construction zones. The VII Coalition, comprised of the U.S. DOT, state DOTs, and automobile manufacturers, estimates the cost to be $5-8 billion to create and $100 million per year to maintain a national VII infrastructure.[15,16]

Moreover, roadway infrastructure is the domain of states and municipalities, which receive funding from the federal government only for select highways. The decentralized nature of highway management can diminish the influence of the federal government in establishing a uniform VII system across the United States.

Technical issues must also be resolved. Particularly, engineers must find a way to implement V2V in vehicles in the near term while ensuring compatibility with VII in the long term. This requires allocating extra computer power for future enhancements of V2V and VII.

VII has increased potential, compared with V2V alone, to address factors in fatal accidents, as shown in Table 1.

Related factor and percent in fatal accidents[10]	Driver aid	+ V2V	+ VII
Failure to keep in proper lane or running off road: 28.5%	LDW, LKA	•	•
Driving too fast for conditions or in excess of posted speed limit or racing: 21.3%		•	•
Under the influence of alcohol, drugs, or medication: 12.7%		•	•
Inattentive (talking, eating, etc.): 7.9%	LDW, FCW		
Failure to yield right of way: 7.3%	BSW	•	•
Overcorrecting/oversteering: 4.6%			
Failure to obey traffic signs, signals, or officer: 4.2%		•	•
Swerving or avoiding due to wind, slippery surface, vehicle/object/person in road, etc.: 3.7%		•	•
Operating vehicle in erratic, reckless, careless, or negligent manner: 3.6%		•	•
Vision obscured (rain, snow, glare, lights, building, trees, etc.): 2.7%	ACC	•	•

Table 1: Safety potential of driver aids, V2V, and VII to affect the top 10 fatal accident factors

HUMAN FACTORS CHALLENGES

The automation of vehicles will likely be an evolutionary process. The market is already seeing automated collision warning and lane departure warning systems leading to automated control-assist devices like active cruise control and lane keep assist. In the distant future, this process will lead to full automation.

DRIVER ROLE CONCERN - Between now (warning and control-assist) and the distant future (full automation), difficulties will arise in clearly defining the driver's role and in assuring the driver understands a vehicle's capabilities. Both of these aspects can adversely impact safety.

In fact, automakers have already begun to diverge in their vehicles' capabilities. There is concern about how a driver who is familiar with one type of system can adapt to driving a vehicle with another, similarly-named system.

Furthermore, in light of increasing automation, experts are considering the likelihood that the driver could lose focus on the driving task.

Stakeholders will need to address these issues as a system. As a 2001 California PATH report states, "the roles of the driver and the automation system will need to be defined so that, when combined, all of the essential safety-critical functions are performed at least as well as they are today." [8]

REMOVING THE DRIVER - By definition, human factors issues can be eliminated by completely removing the driver from the system. Although this would be a worthy goal to pursue, the Department of Defense is taking the lead in seeking autonomous (driverless) vehicles; its aim is to decrease battlefield fatalities. The resulting technology should benefit civilians as well as military personnel.

Specifically, the Defense Advanced Research Projects Agency (DARPA) held racing competitions in 2004, 2005, and 2007 to encourage industry and academic groups to develop autonomous (driverless) vehicles for the military to use in place of conventional vehicles in high-danger situations. In 2007, the race was held for the first time in an urban environment. In the DARPA Urban Challenge, contestant vehicles "simulated military supply missions while merging into moving traffic, navigating traffic circles, negotiating busy intersections, and avoiding obstacles." [17]

The number of finishers for each year's competition showcases the rapid development of fully automated vehicles:

- The 2004 Grand Challenge, held on a 142-mile desert course, finished 0 of 15 finalists.
- The 2005 Grand Challenge, held on a 132-mile desert course, finished 4 of 23 finalists.
- The 2007 Urban Challenge, held on a 60-mile mock-urban course, finished 6 of 11 finalists.[17]

Many experts consider full automation the only way to remove human error from the driving experience. Although the progress made by science and engineering in this area is impressive, most academic, industry, and government experts in attendance for the 2008 IEEE Intelligent Vehicles Symposium predicted that fully-automated, mass-market vehicles would only be available only after 2030, 2040, or later.[18]

MARKET ACCEPTANCE CONCERN - There exists a valid concern over how consumers will accept driver aids, V2V, and VII as on-board helpers to the task of driving. California and Michigan are two states where real-world intelligent vehicle tests involving human participants have been conducted. The participants' feedback about the experience is valuable in determining how the public will react to intelligent vehicles.

California PATH platoon demonstration - An eight-vehicle platoon was demonstrated in 1997 on freeways in San Diego, California. In this demonstration, riders were driven by fully-automated vehicles, and the vehicles safely maintained short separation. Despite the "tailgating" effect of the 21-foot separation that made riders uneasy at first, "most of them quickly adapt and develop a sense of comfort and security because of the constantly maintained separation." [19]

UMTRI pilot test - In a 2008 University of Michigan Transportation Research Institute (UMTRI) report for the DOT, 18 subjects drove vehicles equipped with several driver aids like FCW, LDW, BSW in a pilot test. Following the test, the subjects were asked to subjectively rate the systems on a +2 to -2 scale. The report states, "The mean usefulness score is 1.33 and the mean satisfaction score is 0.75, both of which indicate positive feelings towards [driver aids]."[20]

As part of the same study, UMTRI performed a pilot test for heavy trucks equipped with similar driver aids. Following the test, three of the five truck drivers generally liked the systems, while two of the five disliked the systems due to false alarms. The rate of false alarms will steadily decrease with further development, so this issue should not be a problem with future driver aids.[20]

V2V and VII systems will have much more data to process than current driver aids, so avoiding false alarms will remain an important issue throughout development. Fortunately, many of the driver alerts generated by V2V and VII can be routed through the same user interface as driver aids, which the subjects of the UMTRI study liked.

PUBLIC POLICY ROLE

The U.S. government has the potential to positively impact the development and implementation of intelligent vehicle technologies. Through mandates, rules and regulations, tax incentives or penalties, and subsidies, government has a wide array of options to affect intelligent vehicle technology.

CURRENT INVOLVEMENT - Federal, state, and local government has played a role in several stages of development, from research to field trials, of intelligent vehicle technologies.

Research - The federal government has assisted industry in high-risk research. In a 2008 interview of the administrator for the DOT Research and Innovative Technology Administration, Paul Brubaker, by ITS International, he said, "Government should support basic and applied research, then get...out of the way and let the private sector and localities get on and do things." [14]

Prize competitions - Programs such as the DARPA Urban Challenge for initiating development in automation technology are mutually beneficial to government, industry, and academia. Besides the obvious recognition and awards that winning teams get, government gains an advantage in defense technology, and industry gains know-how and a skilled pool of potential new-hires from academia. Industry also gains solid reassurance that the government is committed to purchasing products and services stemming from this technology in the future.

Technology transfer - In cases where government groups and private industry seek some of the same capabilities, but where each has previously conducted independent development, it can be prudent to exchange knowledge between interested organizations.

Technology transfer has already occurred in the realm of intelligent vehicles. The U.S. Department of Defense, the Department of Transportation, and the Department of Commerce have held a Joint Military/Civilian Seminar On Intelligent Vehicle Technology Transfer. The event has been unclassified and open to all interested parties. At the third, 2008 seminar, leaders of various industry and government projects gave 20 presentations over a two-day period.

Field trials - The federal, state, and local governments have a track record of sponsoring field operation test of near-term advanced technology.

From 2006 to 2008, the VII Coalition, including the U.S. DOT, Michigan DOT, and Oakland County Road Commission, has created the Developmental Test Environment (DTE) in Detroit, Michigan. The DTE demonstrates the proof-of-concept of VII, and 57 sites in Oakland County have been equipped with roadside equipment to communicate with vehicles over DSRC. The DTE will prove the technical viability of the VII system architecture; the DTE also will prove the applications viability of VII to support safety, mobility, and private/commercial services. If the results of DTE are satisfying to industry, they may begin to incorporate VII into their future product plans. The Michigan DOT is also using the DTE to prove operations aspects of VII, such as snow removal and road maintenance.[16]

At the state level, the Michigan Economic Development Corporation is sponsoring the Connected Vehicle Proving Center along with several industry partners. The center will allow developers to share costs and coordinate testing in expensive facilities and in public roadways. Similar joint efforts between federal, state, and local agencies are taking place in 13 other states across the U.S.[16]

STANDARDS DEVELOPMENT - A major challenge in developing a vast new intelligent transportation system is getting agreement from disparate parties in the automobile industry, academic institutions, and from state and federal government research agencies and regulatory bodies.

The government, by executive decree, is obligated to use voluntary consensus standards developed by the private sector whenever practicable. According to the Executive Office of Management and Budget in its Circular No. A-119, this applies to all agencies of the federal government. Although voluntary consensus means that all parties will be in agreement, this approach can take much longer than mandates used in other countries.[21]

REWARDING IMPLEMENTATION - Implementing new technologies on a wide scale can entail high initial costs that can make them unattractive to consumers. Government can stimulate sales by providing incentives in the interim deployment stage, until the technology becomes established.

Performance-based incentives - Government could carefully create performance criteria for awarding tax credits or subsidies to the customer.

For driver aids alone, the federal government could subsidize the cost of safety features like CACC, LDW, and FCW. The incentive would be vehicle-specific and be based on the capability of its driver aids. In cases where vehicle manufacturers offer safety features bundled with luxury amenities, government could impose a rule to separate the safety options from amenities like leather seats and entertainment systems.

For V2V, the government could financially assist those who seek to retrofit their vehicle with DSRC or those who purchase a new vehicle equipped with DSRC. As a comparable precedent, the government-mandated 2009 switch from analog to digital television comes with a subsidy. Called the TV Converter Box Coupon Program, it allows all U.S. households to receive two $40 coupons toward a digital-to-analog converter for old, analog television sets (these converters cost approximately $60). Similarly, government could offer coupons for retrofitting vehicles with DSRC transceivers while mandating that new vehicles come equipped with them.[22]

Another example is the Federal government's New Energy Tax Credits for Hybrids, which varies the tax credit according to the mileage performance of the hybrid vehicle and phases out the incentive after sales of a model reach 60,000 units. Similarly, the incentive could be greater for more capable V2V systems than for less capable systems, and the incentive could be gradually phased out once a "critical mass" of V2V-equipped vehicles are sold. After a certain portion of the market possesses the technology, economies of scale and desire to offer competitive features could drive down cost and thereby increase market penetration.

For VII, the customer is not the car-buying public, but rather the state or local government that is considering an infrastructure upgrade to adopt the VII standard. Performance-based incentives could depend on the safety, congestion, and environmental effects of the planned infrastructure over the current condition. This would allow states and local governments the freedom to decide which roads to equip with VII first.

Marketing strategies - For all driver aids, V2V, and VII, the federal government can provide public awareness of what the safety technologies mean for them.

Crash *avoidance* ratings - Through NHTSA's New Car Assessment Program (NCAP). NCAP has provided star ratings (based on a five-star scale) for front and side impacts. These ratings, along with the Insurance Institute for Highway Safety (IIHS) crash test ratings, have influenced consumer purchasing habits.

More recently, NHTSA has begun evaluating and assigning stars for roll-over safety. In addition to awards for a vehicle's crashworthiness, NHTSA and IIHS could evaluate the vehicle's performance in avoiding crashes altogether. Since driver aids are available now, NHTSA and IIHS could begin awarding stars and ratings for today's technology, and then increase requirements for high ratings as new technologies emerge.

Addition to fuel economy ratings - Since V2V and VII also affect fuel economy of the equipped car as well as having a multiplier effect on other equipped- and non-equipped vehicles, the EPA could add a numerical value of the fuel savings next to the standard miles-per-gallon rating. For example, if a 40-mpg vehicle lowers consumption by 10% when using CACC, its new EPA fuel economy could read "40 mpg + 4 Intelligent Vehicle mpg)."

ACCOUNTING FOR UNINTENDED CONSEQUENCES - Given the potential benefits of V2V and VII systems, it may be easy to forget to consider the side effects caused by their implementation. Careful government action can help to mitigate the negative effects on the state and local levels.

Increased vehicle use and tolling - Advancements in fuel economy standards and reduction of congestion might influence vehicle use in relation to other modes of transportation. However, past CAFE increases have not significantly decreased overall fuel consumption in the U.S., because it lowered transportation fuel costs and subsequently increased VMT. Taking this history into account, 35 mpg by 2020 may bring with it the unintended consequence of increasing travel and *not* achieving its intended goal of decreasing overall consumption.

VII holds one possible answer in open road tolling. State and local governments can enact usage fees for driving on the most heavily congested roads, using DSRC for collection at cruising speed. The driver would pay for exactly the amount of road driven. The proceeds from these tolls would be ideally suited for implementing more VII capability around the state or municipality, in effect creating a self-perpetuating system after an initial investment in heavily congested areas.

The federal government grants authorization for states to enact tolling on Interstate highways. However, much of the public often opposes such measures, because tolls become revenue-generating sources for other state spending purposes.

Therefore, careful stewardship of these tolls would be required to ensure that they reflect the cost of the infrastructure, and not other government programs. This can be ensured by enacting policies that limit the use of toll income to further investments in road infrastructure.

Safety problems and investigation - Many lessons have been learned following the ill-fated introductions of automatic seat belts and high-powered first-generation airbags, and engineers are now thoroughly testing every piece of technology that goes onto a vehicle, especially safety equipment. However, there is a chance that a critical algorithm, component, or safeguard will be overlooked. For this reason, NHTSA has a complaints database that can be accessed by concerned members of the public.

If an accident of national importance does occur, the National Transportation Safety Board (NTSB) has the mandate to perform a thorough and impartial investigation and recommend actions to the appropriate organs of government and industry. However, it can take many years for safety problems to manifest themselves, and to do so generally requires extensive market penetration of the problematic technology. Needless to say, this means of addressing safety issues reactively is the least desirable option. Also, it can take days to years for NTSB recommendations to be implemented by the responsible party.

CONCLUSION AND RECOMMENDATIONS

Among new transportation technologies, intelligent vehicles provide an attractive mix of benefits to safety, congestion, fuel efficiency, and environmental friendliness. The improvements afforded by current, near-term, and long-term intelligent vehicle technologies are real, and they are being proven by current pilot programs. Challenges introduced by increased automation and complex systems integration are being resolved by engineers around the world.

However, the success of intelligent vehicle technologies ultimately depends on the actions of a few—and often non-unified—key players in the public policy arena. Fortunately, it is the public who choose (indirectly, in some cases) the policymakers. It is our duty as citizens to make sure our voices are loud enough, and it is the duty of the policymakers to listen.

For a timely and efficient transition to intelligent vehicle technologies, government should do the following:

- Continue to sponsor intelligent vehicle competitions, technology transfer, and field trials
- Add to NHTSA star ratings to reflect the active safety benefits of V2V—the deployment of which should coincide with the first launch of the technology
- Add to EPA fuel economy ratings to reflect the energy savings of V2V—the deployment of which should coincide with the first launch of the technology
- Offer performance-based incentives to car buyers during the introduction of V2V by offering subsidies for retrofitting DSRC to existing vehicles or for buying V2V-equipped new vehicles.
- Offer performance-based federal incentives and disincentives to state and local governments during the implementation of a standardized VII system
- Approve state and local governments to use VII-based toll collection on Interstate highways, with funding restricted to roadway projects and further VII implementation.

ACKNOWLEDGMENTS

The author wishes to thank the following people for their wise advice and invaluable assistance: Mr. Timothy Mellon, SAE International Government Affairs Director; Dr. Jeffrey King, 2008 WISE Faculty Member in Residence and Assistant Professor at the Missouri University of Science and Technology; Mr. James Bryce, 2008 WISE Intern, University of Missouri; and Ms. Brooke Buikema, 2008 WISE Intern, Calvin College.

REFERENCES

1. "Highway Statistics," HM-60 and VM-2, Federal Highway Administration, Washington, DC, 1997-2007.
2. Boeing Commercial Airplanes - Boeing 747-8, http://www.boeing.com/commercial/747family/747-8_facts.html, accessed July 25, 2008.
3. "Traffic Safety Facts, 2006 Data: Overview," pp. 1-3, DOT HS 810 809, National Highway Transportation Safety Administration, Washington, DC, 2008.
4. Schrank, David, Tim Lomax, "The 2007 Urban Mobility Report," Texas Transportation Institute, http://mobility.tamu.edu, 2007.
5. "Greenhouse Gas Emissions from the U.S. Transportation Sector, 1990–2003," pp. 8-11, EPA 420 R 06 003, United States Environmental Protection Agency, Washington, DC, 2006.
6. "Energy Independence and Security Act of 2007," Title I, Public Law 110-140, United States of America, 2007.

7. Davis, Stacey, "Transportation Energy Data Book," p. "7-22," Oak Ridge National Laboratory, National Technical Information Service, Springfield, VA, 2001.
8. Shladover, Steven, et al, "Development and Performance Evaluation of AVCSS Deployment Sequences to Advance from Today's Driving Environment to Full Automation," pp. 72-77, California Partners for Advanced Transit and Highways, Berkeley, CA, 2001.
9. Jiang, Daniel and Luca Delgrossi, "IEEE 802.11p: Towards an International Standard for Wireless Access in Vehicular Environments," pp. 2036-2040, Vehicular Technology Conference, 2008, IEEE, Singapore, 2008.
10. "FARS/GES 2006 Data Summary," p. 19, DOT HS 810 819, National Highway Transportation Safety Administration, Washington, DC, 2008.
11. van Arem, Bart, et al, "Design and evaluation of an Integrated Full-Range Speed Assistant," p. 44, TNO Traffic and Transport, The Netherlands, 2007.
12. "21st Century Truck Partnership: Roadmap and Technical White Papers," p. 2, 21CTP-0003, Department of Energy, Washington, DC, 2006.
13. Hellström, Erik, Anders Fröberg, Lars Nielsen, "A Real-Time, Fuel-Optimal Cruise Controller for Heavy Trucks Using Road Topography Information," 04-03-2006, SAE International, Warrendale, Pennsylvania, 2006.
14. "Driven Man," pp. 16-17, ITS International magazine, January/February 2008.
15. Meeting with Raj Ghaman, Travel Management Team Leader, Federal Highway Administration Office of Operations R&D, July 10, 2008.
16. VII Coalition, http://www.vehicle-infrastructure.org.
17. DARPA Urban Challenge, http://www.darpa.mil/grandchallenge/, Accessed July 20, 2008.
18. E-mail communication with Steven Shladover, California PATH and Keynote Speaker in IEEE Intelligent Vehicles 2008 Symposium, June 30, 2008.
19. "Vehicle Platooning and Automated Highways," California Partners for Advanced Transit and Highways, http://www.path.berkeley.edu/PATH/Publications/Media/FactSheet/VPlatooning.pdf.
20. Green, P. J.Sullivan, O. Tsimhoni, J. Oberholtzer, M.L. Buonarosa, J. Devonshire, J. Schweitzer, E. Baragar, J. Sayer, "Integrated Vehicle-Based Safety Systems (IVBSS): Human Factors And Driver-Vehicle Interface (DVI)Summary Report," pp. 86-93, DOT HS 810 905, U.S. Department of Transportation, Washington, DC, 2008.
21. Circular No. A-119, Executive Office of Management and Budget, http://www.whitehouse.gov/omb/circulars/a119/a119.html.
22. TV Converter Box Coupon Program, https://www.dtv2009.gov.

CONTACT

Eric Sauck is pursuing a Bachelor of Science in Mechanical Engineering and a Master of Automotive Engineering, both at the University of Michigan. He has worked for several automotive suppliers and manufacturers. In the summer of 2008, he was sponsored by SAE International to participate in Washington Internships for Students of Engineering (www.wise-intern.org) to gain exposure to the public policy process.

Questions, comments, and discussion should be directed to esauck@umich.edu.

DEFINITIONS, ACRONYMS, ABBREVIATIONS

Light-duty vehicle: Light-duty vehicles are defined as vehicles with a gross vehicle weight rating (GVWR) of less than 8,500 lbs. They include passenger cars, sport-utility vehicles, minivans, pickup trucks, and motorcycles.

ACC: Active cruise control

BSW: Blind Spot Warning

CACC: Cooperative active cruise control

DARPA: Defense Advanced Research Projects Agency

DOT: Department of Transportation

DSRC: Dedicated short range communications

EPA: U.S. Environmental Protection Agency

FCW: Forward collision warning

GHG: Greenhouse gas

GPS: Global Positioning System

IEEE: Institute of Electrical and Electronics Engineers

IIHS: Insurance Institute for Highway Safety

LKA: Lane keep assist

LDW: Lane departure warning or lateral drift warning

mpg: Miles per gallon

NCAP: National Highway Transportation Safety Administration New Car Assessment Program

NHTSA: National Highway Transportation Safety Administration

PATH: California Partners for Advanced Transit and Highways

TTI: Texas Transportation Institute

V2V: Vehicle-to-vehicle communications

2009-01-0165

Prioritized CSMA Protocol for Roadside-to-Vehicle and Vehicle-to-Vehicle Communication Systems

Jun Kosai, Shugo Kato, Toshiya Saito, Kazuoki Matsugatani and Hideaki Nanba
DENSO CORPORATION

ABSTRACT

This paper proposes Prioritized-CSMA (Carrier Sense Multiple Access) protocol for Japanese vehicle safety communications (VSC). To realize Japanese VSC, we have studied a protocol to carry out Roadside-to-Vehicle (R2V) and Vehicle-to-Vehicle (V2V) communications on single channel because a single 10MHz bandwidth channel on UHF band is allocated for VSC in Japan. In this case, R2V communication requires higher quality than V2V communication, so we have developed a protocol to prevent interference between R2V and V2V communications. The proposed protocol has been evaluated by field experiments and a simulation. The results confirm that the proposed protocol prevents the interference effectively and it has capability to achieve high quality R2V communication in actual case.

INTRODUCTION

Vehicle safety communications (VSC) including Roadside-to-Vehicle (R2V) and Vehicle-to-Vehicle (V2V) have been studied to improve vehicle safety. To realize VSC, frequency channels are allocated, and communication standards have been developed all over the world. In US and EU, multiple frequency channels on 5.9GHz band are allocated to VSC. From the viewpoint of communication standard, IEEE802.11p and IEEE1609 are used in US, and the same or similar standard would be used in EU[1-4]. On the other hand in Japan, a single 10MHz bandwidth channel on UHF band is allocated to VSC, but the standard for these communications is still being discussed. Japanese VSC environment differs from US and EU, so a communication protocol suitable for Japanese VSC is required.

Japanese VSC has an important issue to develop communication protocol. The issue is that both R2V and V2V communications have to be carried out on allocated single channel because allocated channel is just 10MHz single channel. From the viewpoint of communication requirements, R2V communication takes priority over V2V communication because roadside units (RSU) can send packet including crucial information to avoid traffic accidents especially at the intersection. However, in this case where both of communications are on single channel, interference between R2V and V2V should occur and affect the performance each other. Especially interference from V2V to R2V should be critical problem.

To prevent the interference, this paper proposes Prioritized-CSMA protocol, which is a CSMA-based medium access control (MAC) protocol. We assume that CSMA-based protocol is suitable for VSC because on-board units (OBU) can send their own packets without centralized scheme. To adapt CSMA protocol to Japanese VSC, we introduce a concept of time division into R2V and V2V time slots. RSUs play as a master that controls R2V and V2V slot timings; OBUs refrain from sending their packets during R2V slot and transmit packets during V2V slot by CSMA scheme. To evaluate the proposed protocol, we have developed UHF band wireless communication units and performed field experiments. Furthermore, we have evaluated the protocol in a scenario where large amount of vehicles exist by simulation.

This paper is organized as follows: Second section explains Japanese VSC system we assume. Third section presents our proposed protocol for Japanese VSC. Fourth section evaluates the proposed protocol by

The Engineering Meetings Board has approved this paper for publication. It has successfully completed SAE's peer review process under the supervision of the session organizer. This process requires a minimum of three (3) reviews by industry experts.

ISSN 0148-7191

9-2009-01-0165

SAE International

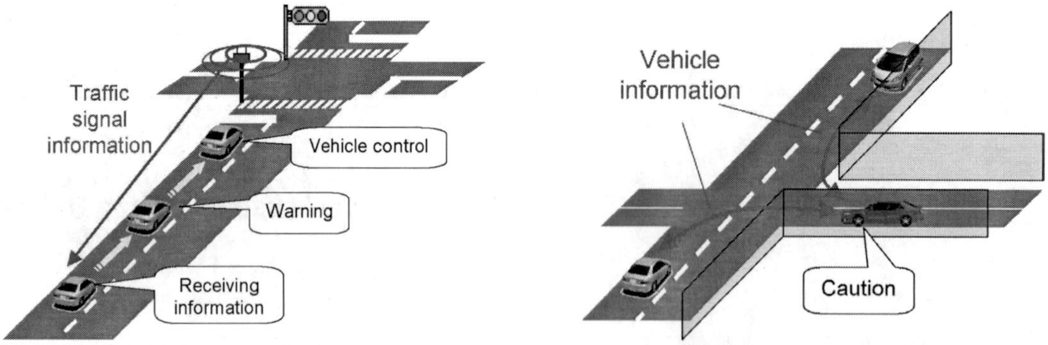

(a) R2V type application (b) V2V type application
Figure 1 Examples of Japanese VSC application

field experiments and a simulation. The last section describes the conclusion.

JAPANESE VSC SYSTEM

Vehicle safety applications and VSC systems have been discussed and developed by various organizations[5-7], and it is difficult to define Japanese VSC clearly. Therefore, this section introduces our assumed Japanese VSC system from the viewpoint of applications and requirements.

APPLICATIONS - Japanese VSC contains applications using both R2V and V2V communications. Figure 1 shows the examples. Figure 1 (a) depicts an example of driving support system based on roadside information. RSU sends traffic signal information and OBU, which receives the information, performs warning or vehicle control according to the situation. To realize vehicle control by roadside information, high communication quality is required. Figure 1 (b) illustrates an example of driving support system based on V2V communication. Each vehicle broadcasts the own data including the position, speed and heading. A receiving vehicle provides the driver with approaching vehicle information.

REQUIREMENTS - We assume that following four requirements are important to develop communication protocol for Japanese VSC.

1. Both RSU and OBU send packets on the same channel.
2. R2V communication has priority over V2V.
3. Both RSU and OBU broadcast their packets.
4. R2V communication occupancy is adaptable to R2V data amount.

Firstly, as explained above, a single channel on UHF band is allocated for VSC in Japan; RSU and OBU have to share the same channel.

Secondly, we assume that RSU has much more information than OBU because RSU would be connect with some facilities as traffic information center, database and so on via wired or wireless connection. Therefore RSU would transmit packets including more precise and crucial information to avoid traffic accidents especially at the intersection. That's why RSU transmission should not be interfered from OBU transmission. This paper aims to obtain 99% or more packet arrival rate for R2V communication.

Thirdly, every communication units send the information by broadcast. Unicast transmission is not suitable for safety application because a sending node cannot specify the receiving node in advance. We focus on broadcast transmission only. In addition in this paper, R2V communication means that RSU only sends packets and gives information to OBUs unilaterally.

Lastly, flexibility of R2V communication occupancy is required. The size of data, which is transmitted from RSU, may be changed depending on the situation such as supplying application, RSU located position and so on. Moreover, if there is no RSU, all communication resources should be assigned for OBU. Therefore, this is important requisite for Japanese VSC.

PRIORITIZED-CSMA PROTOCOL

This section proposes Prioritized-CSMA (P-CSMA) protocol. As mentioned above, we have developed a new MAC protocol based on CSMA because CSMA protocol does not need centralized scheme. In the case where CSMA protocol is applied to Japanese VSC, however, it is difficult to achieve 99% packet arrival rate for R2V communication caused by interference from V2V communication. This section states an issue, which is caused by applying CSMA protocol and presents P-CSMA as the solution.

INTERFERENCE FROM V2V - CSMA protocol has well-known problem, hidden node problem[8]. Figure 2 shows a situation where the hidden node problem occurs between R2V and V2V communications. While RSU sends a packet to OBUs within the communication range, the outside OBUs would send a packet simultaneously. As the result, packet collision happens between R2V and V2V. This collision leads to degradation of packet arrival rate for R2V communication.

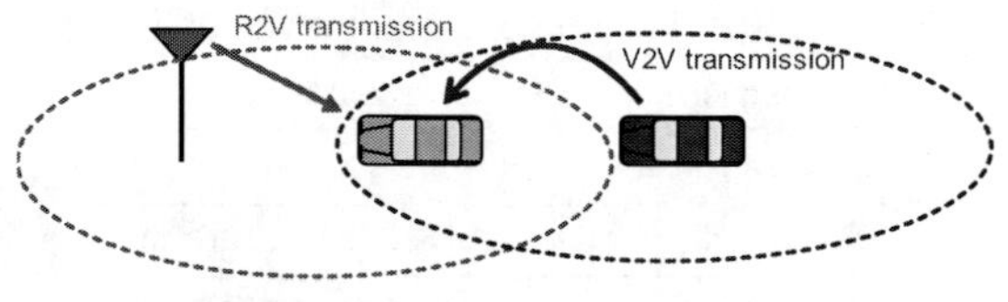

Figure 2 Hidden node problem

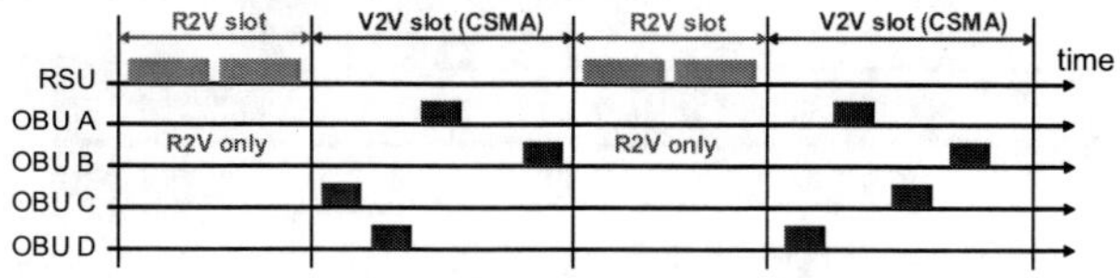

Figure 3 Division of R2V slot and V2V slot

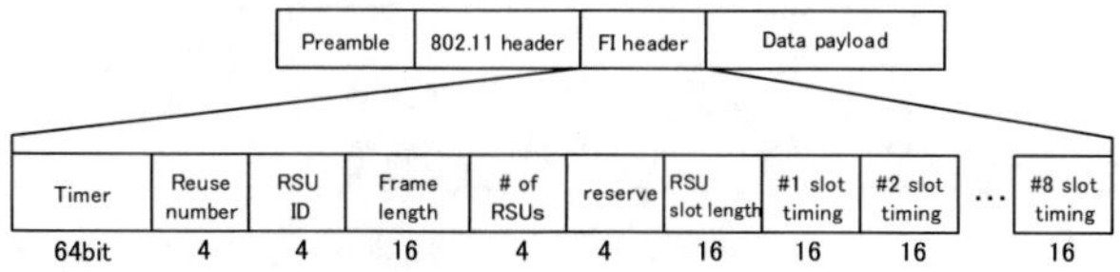

Figure 4 Frame format

PROTOCOL DESIGN

Concept - To avoid interference between R2V and V2V, we introduce time division scheme into ordinary CSMA. Time period is divided into two parts, R2V slot and V2V slot. Here, slot means a period that is allocated to RSU or OBU. Figure 3 shows the concept of P-CSMA. RSU works as a master to control R2V and V2V slots, and the slots are allocated by RSU. RSU sends packets only in R2V slot and OBU transmission is prohibited during the R2V slot. On the other hand in V2V slot, only OBU sends packets using conventional CSMA scheme. In this way, R2V communication is protected against the interference from V2V. In addition, R2V slot length can be adapted to R2V data amount because RSU assigns the slot itself.

To realize the concept, following two requirements have to be satisfied.

1. RSU informs OBU of slot information (SI).
2. OBU, which is hidden node from RSU, has to know the SI.

Hereinafter, this section describes frame format and SI propagation to meet the requirements.

Frame format - RSU sends SI to OBU to assign R2V and V2V slots. The SI includes *Timer*, *Slot length*, *Slot timing*. Figure 4 illustrates frame format and Table 1 shows the contents. SI is inserted between IEEE802.11 header and data payload. *Timer* is cyclic time in microsecond, and it is used for time synchronization between RSU and OBU. *Reuse Number (RN)* indicates freshness of the SI. The *RN* is utilized for updating SI on OBU and managing SI propagation area.

Table 1 Field contents of slot information

Field	Contents
Timer	Reference time which RSU keeps and generate using GPS.
Reuse Number	Indicator for freshness of slot information.
RSU ID	Identification of RSU.
Slot cycle	Slot cycle in which RSU send packets.
# of RSUs	The number of RSU slot timing in the slot information.
RSU slot length	RSU slot length calculated from RSU data amount.
#x slot timing	Start timings of RSU slot. (This field can contain up to 8 slot timings)

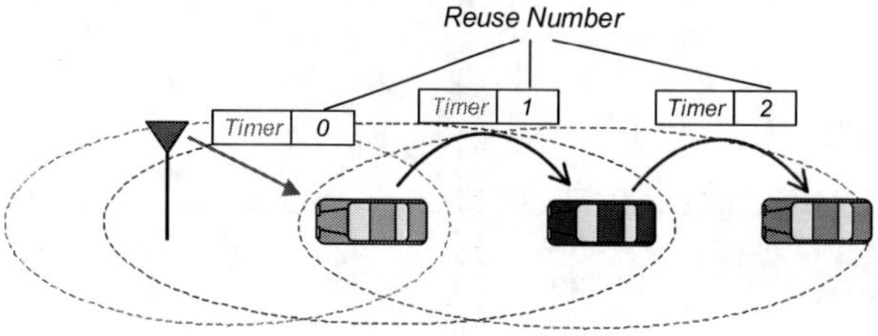

Figure 5 Propagation of slot information

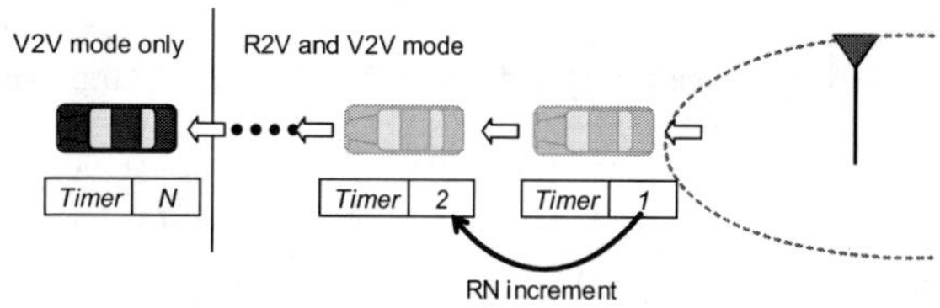

Figure 6 RN increment by elapsed time

SI is capable of including multiple *slot timings* (up to 8). As a result, P-CSMA works properly by containing all slot timing of neighbor RSUs even if multiple RSUs exist in the neighborhood.

Propagation of slot information - SI have to reach OBUs which are hidden nodes from RSU. A RSU generates SI and sends the SI to OBUs within the RSU communication range; however, the SI cannot reach to the hidden nodes. Therefore, OBU retransmits the received SI to surrounding OBUs.

Figure 5 shows the SI propagation. OBU A, which have received SI from RSU directly, synchronizes own timer with RSU using *Timer* field and controls to refrain from sending its own packets during R2V slot referring to *RSU slot length* and *slot timing* fields. When OBU A sends its own data packet, OBU A sends the packet in the same format as Figure 4. At this moment, OBU A stamps its own timer value on *Timer* field, and the other field are copied from received SI. In the same way, OBU B retransmits SI after receiving a packet from OBU A. In this way, SI is delivered to hidden nodes from RSU.

However, SI propagation should be limited within appropriate area. If OBUs, which do not interfere with

Figure 7 Experimental communication unit

R2V communication, receive SI, the prohibition of transmission causes unnecessary restrictions on V2V communication. So, P-CSMA controls the SI propagation area by *RN,* which indicates proximity to RSU and freshness of SI. SI is discarded when *RN* is over threshold. A couple of *RN* control methods are explained as follows.

<u>SI retransmission</u> - *RN* in SI is incremented at every SI retransmission as shown in Figure 5. When RSU sends SI, RN is 0, and the *RN* is incremented when OBU receives the SI. For example, *RN* is 3 on OBU C in Figure 5. If the *RN* threshold is 3, OBU C does not retransmit SI to surrounding OBUs. In this way, SI propagation area is controlled.

<u>Elapsed time since latest SI</u> - The other method to control *RN* uses elapsed time since receiving latest SI. Figure 6 shows the mechanism. OBU cannot receive SI when OBU get out of RSU communication range, but if OBU continues to keep the SI, V2V communication is restricted uselessly. To avoid this problem, *RN* is incremented every time definite period is elapsed. When *RN* is over the threshold, SI is invalidated. After that, OBU shifts to V2V only mode, in which OBU can send packets anytime.

P-CSMA uses both of methods to control SI propagation area.

EXPERIMENTAL COMMUNICATION UNIT - We have developed experimental communication units for performance evaluation. Figure 7 shows the external appearance, and Table 2 shows the specification. Slot cycle means intervals at which RSU sends packets. Now slot cycle is 100 ms; if 20 % are allocated for R2V, 20 ms are used for R2V and 80ms are used for V2V. To realize high accuracy time synchronization, the synchronization function is implemented in hardware logical circuit. As the result, the synchronization accuracy is less than 4us, which is sufficient for P-CSMA.

EVALUATION

This section describes performance evaluation of P-CSMA. The evaluation consists of two field experiments and a simulation analysis. Filed experiments are conducted to confirm that P-CSMA is capable of preventing interference from V2V to R2V. In addition, the simulation is carried out for performance analysis in the case where more than hundreds of OBUs exist.

Table 2 Specification of experimental unit

Center frequency	792.5MHz
Maximum TX power	20dBm
Modulation	BPSK, QPSK, 16QAM
MAC protocol	Prioritized-CSMA
Slot cycle	100ms
Synchronization accuracy	less than 4us
Size	330(W) × 200(D) × 80(H)

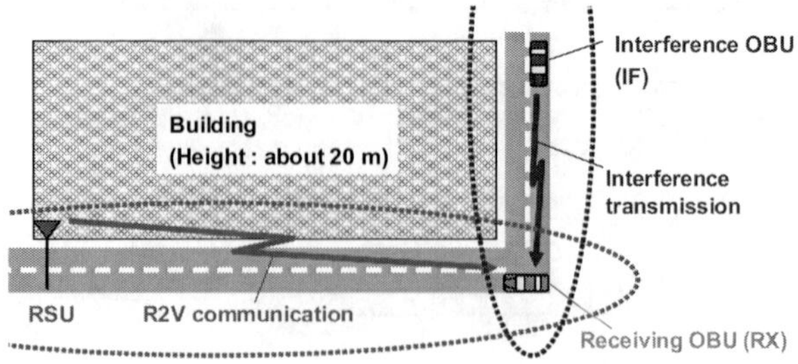

Figure 8 Experimental arrangement in static environment

Table 3 Experimental parameters

Bandwidth		12 Mbps
Slot allocation		R2V:20%, V2V:80%
RSU	Packet size	1500 byte
	Transmission rate	100 packets / 100ms
Receiving OBU	Packet size	100 byte
	Transmission rate	1 packet / 100ms
Interference OBU	Packet size	100 byte
	Transmission rate	0 ~ 100 packets / 100ms

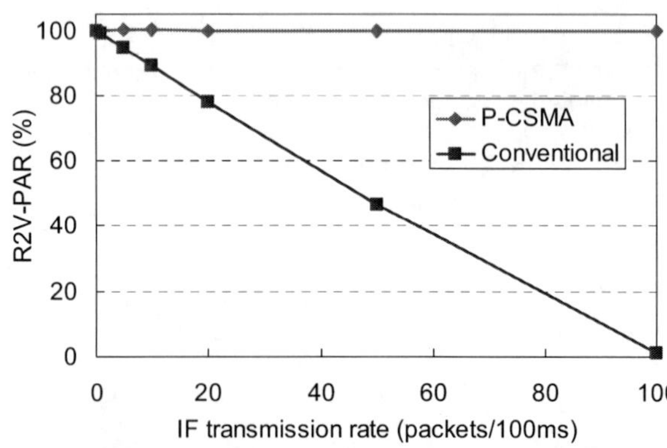

Figure 9 R2V-PAR vs. IF transmissions

FIELD EXPERIMENT IN STATIC ENVIRONMENT - This experiment is intended to evaluate the performance depending on interference from hidden node. Figure 8 depicts the experimental arrangement. Three

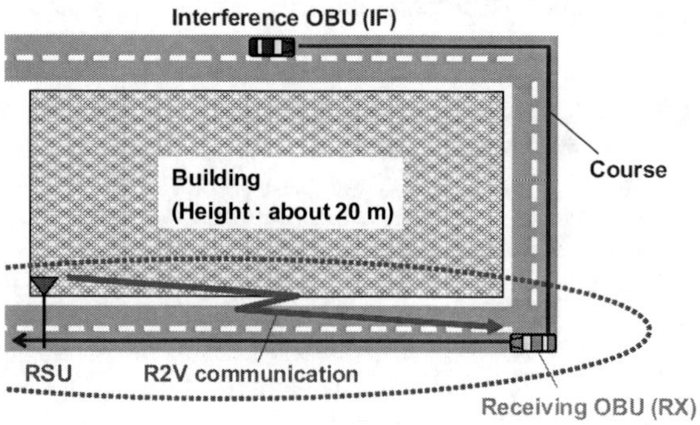

Figure 10 Experimental arrangement in dynamic environment.

communication units are located as RSU, receiving OBU (RX) and interference OBU (IF). Although RX can communicate with both RSU and IF, RSU and IF cannot communicate each other. In other words, RSU and IF are mutually hidden nodes. Therefore, if RSU and IF generate sending packets simultaneously, packet collision happens at RX because of incapable of carrier sense. In this arrangement, we compared R2V packet arrival rate (R2V-PAR) of P-CSMA with conventional CSMA's. Here R2V-PAR means communication successful rate from RSU to RX. Table 3 shows the experimental parameters. Slot allocations for R2V and V2V are 20 % and 80 % respectively. The packet size of RSU and OBU comes from supposed applications. We assume that OBU sends a packet in every 100 ms, so RX sends a packet per 100 ms. IF Transmission rate changes from 0 to 100 packets per 100ms. The number of packets per 100 ms from IF correspondents to the number of hidden nodes from RSU.

Figure 9 presents the experimental result, which is relation between the number of IF packets and R2V-PAR. The result shows that R2V-PAR of conventional method decreases as IF transmission rate increases. The reason of the R2V-PAR degradation is interference from IF to R2V communication. IF cannot sense R2V communication, so R2V-PAR degrades linearly as IF transmission rate increases. On the other hand, P-CSMA keeps almost 100% R2V-PAR regardless IF transmission rate. This result confirms that P-CSMA works properly to prevent hidden node problem on R2V communication in static environment.

FIELD EXPERIMENT IN DYNAMIC ENVIRONMENT - This experiment evaluates the performance in dynamic environment, where OBUs come into RSU communication range from outside. Figure 10 depicts the experimental arrangement. RSU and RX are placed in the same positions as the previous experiment. IF starts from outside of RX communication range and runs along the course shown in Figure 10 toward RSU. The vehicle speed is 20 km/h. IF does not have valid SI initially because IF is placed on outside of RX communication range. Considering this point, this experiment measures an effect of IF movement at the moment when IF receives SI. IF transmission rate is 20 packets per 100 ms.

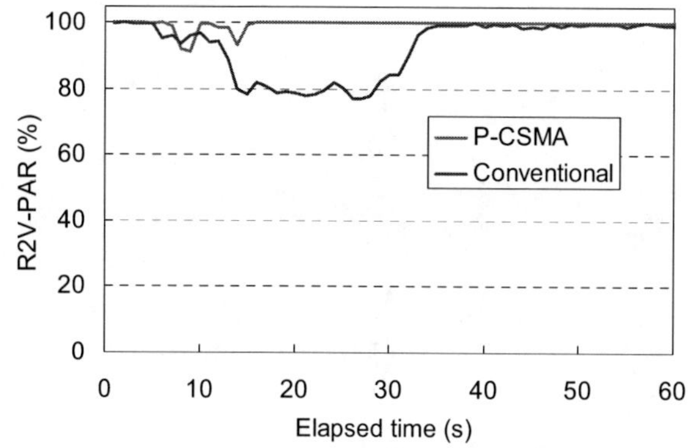

Figure 11 R2V-PAR vs. elapsed time

Figure 11 shows the experimental result. The horizontal axis indicates elapsed time from IF starting. The vertical axis indicates R2V-PAR at every second. In the graph, IF enters RX communication range at around 5 s, and IF enters RSU communication range at around 35 s. In other words, IF is hidden node from RSU between 5 s and 35 s. During the period, R2V-PAR of conventional method degrades to 80% because of hidden node effect. On the contrary, P-CSMA keeps higher R2V-PAR compared with conventional method. This result shows that P-CSMA functions effectively to keep R2V-PAR even in dynamic environment. However, instantaneous R2V-PAR drops at around 10s. Hereinafter, we consider why R2V-PAR drops at the timing.

The R2V-PAR degradation is cased by two reasons. First, a relation between communication range and radio propagation range affects R2V-PAR. Generally, radio propagation range is greater than communication range, where two nodes are able to send and receive packets each other. In this experiment, IF transmission interferes with receiving RSU packets at RX before IF receives SI from RX. Second, the degradation depends on experimental scenario. In this experiment, RX is the only node which can receive RSU packet and retransmit SI. In contrast, the number of hidden nodes from RSU corresponds to 20 because IF sends 20 packets per 100 ms. Therefore, this experimental scenario is the hardest case for P-CSMA, where delivering SI to hidden node is difficult.

When the actual operability of P-CSMA is considered, the result suggests that how long SI can spread is the most important for P-CSMA. So, we calculate the time by following simulation.

SIMULATION - We calculated the time SI spread in traffic flow by using traffic and network simulator. Figure 12 shows the simulation model and Table 4 shows the parameters. The duration SI spreads depends on vehicle density; this simulation evaluates the duration with multiple vehicle densities. From the communication ranges of RSU and OBU, hidden nodes appear in shaded area of Figure 12. So, the percentage of OBUs,

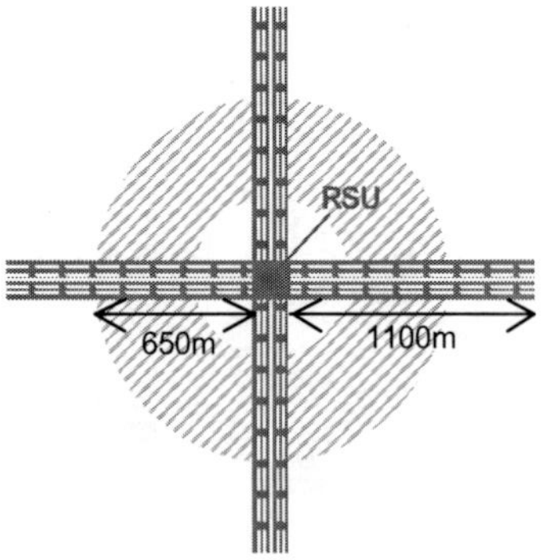

Figure 12 Simulation model

Table 4 Simulation parameters

Lane	Three lanes each way
Vehicle density	5, 10, 20, 30 vehicles/km
RSU communication range	200 m
OBU communication range	450 m
R2V slot rate	5 %

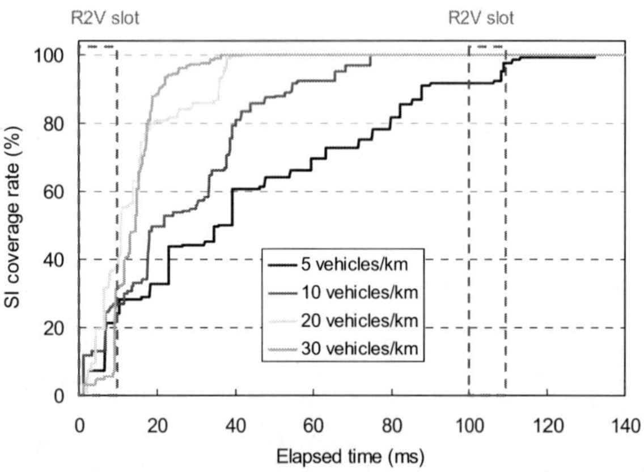

Figure 13 SI coverage rate vs. elapsed time

which have received SI in 650 m radius from RSU, is defined as SI coverage rate.

Figure 13 shows a relation between elapsed time and SI coverage rate. In the case of high vehicle density, SI coverage rate rises rapidly and it reaches 100 % in 40 ms. Also, SI coverage rate increases to 100 % in approximately 100 ms even if vehicle density is 5 vehicles per km.

This result demonstrates that SI can spread out over the area necessary to prevent hidden node problem in short period. Therefore, P-CSMA is effective to realize reliable R2V communication in actual traffic flow.

CONCLUSION

This paper proposed Prioritized-CSMA protocol to realize Japanese VSC, in which both R2V and V2V communications are carried out on single channel. We assumed that R2V communication takes priority over V2V communication, so we developed P-CSMA to prevent interference from V2V to R2V. P-CSMA introduced time slots of R2V and V2V to conventional CSMA by sending SI from RSU.

We developed experimental communication unit to evaluate P-CSMA and conducted field experiments. The experimental results show that P-CSMA improves the R2V-PAR sufficiently compared with conventional method. In addition, the simulation result demonstrates that P-CSMA has capability to achieve high quality R2V communication in actual case because SI coverage rate reaches 100 % in short period.

REFERENCES

1. Wireless LAN Medium Access Control (MAC) and Physical Layer (PHY) Specifications, IEEE Std 802.11-2007, 2007
2. Wireless LAN Medium Access Control (MAC) and Physical Layer (PHY) Specifications Wireless Access in Vehicular Environments, IEEE 802.11p/D3.0, 2007
3. Wireless Access in Vehicular Environments (WAVE) – Networking Services, IEEE Std 1609.3-2007, 2007
4. Wireless Access in Vehicular Environments (WAVE) – Multi-channel Operation, IEEE Std 1609.4-2006, 2006
5. O. Maeshima, S. Cai, T. Honda, and H. Urayama, A roadside to vehicle communication system for vehicle safety using dual frequency channels, presented at IEEE ITSC 2007, pp.349-354, 2007
6. Y. Tadokoro, K. Ito, J. Imai, N. Suzuki, and N. Itoh, A New Approach for Evaluation of Vehicle Safety Communications with Decentralized TDMA-based MAC Protocol, presented at V2VCOM2008, 2008
7. K. Tokuda, M. Akiyama, H. Fujii, DOLPHIN for inter-vehicle communications system, presented at IV2000, pp.504-509, 2000
8. F. A. Tobagi and L. Kleinrock, (1975). Packet Switching in Radio Channels: Part II – the hidden terminal problem in carrier sense multiple-access modes and the busy tone solution, IEEE Trans. in Communications, COM-23 (12), pp.1417-1433

2009-01-0163

Communication in Future Vehicle Cooperative Safety Systems: 5.9 GHz DSRC Non-Line-of-Sight Field Testing

Radovan Miucic and Tom Schaffnit
Honda R&D Americas, Inc.

ABSTRACT

Dedicated Short Range Communication (DSRC) is increasingly being recognized as the protocol of choice for vehicle safety applications by Original Equipment Manufacturers (OEMs) and road operators. DSRC offers the ability to communicate effectively from vehicle-to-vehicle and from vehicle to infrastructure with low latency and high reliability. A wide range of applications have been conceptualized to support safety, mobility and convenience, including: cooperative collision avoidance, travel information, and electronic payment. To be effective, infrastructure-based applications require an installed-vehicle base along with infrastructure deployment, while vehicle-to-vehicle applications require significant DSRC market penetration along with some degree of infrastructure support systems. Some vehicles currently include safety applications involving forward looking radar. The radar supplies information about objects, their distances and relative speed ahead of the host vehicle. When LIDAR (Light Detection and Ranging) becomes feasible for automotive usage, it may also offer an alternative digital vision of the objects in front of the host vehicle. Camera vision is another option for sensing surrounding vehicles. Radar, LIDAR, and camera have advantages and limitations in terms of range and directionality. OEMs are investigating DSRC as a means to enhance visibility of the oncoming and surrounding traffic, particularly in places where line-of-sight is obstructed by other vehicles, buildings, corners etc. Future cooperative vehicle safety applications are expected to be mainly communication-based. A significant challenge lies ahead in combining and processing enormous amounts of information from the host, surrounding vehicles and infrastructure in real time fashion. This paper will focus on the expected potential of such visibility enhancement through the use of DSRC communication.

INTRODUCTION

In the last ten years, many automotive OEMs, their tier 1 suppliers, research universities and wireless radio manufacturers, have been investigating use of DSRC for vehicle-to-vehicle and vehicle-to-infrastructure applications [1] [2] [3] [4] [5] [6]. The major U.S. Department of Transportation (USDOT) initiative called "Vehicle Infrastructure Integration" (VII) [7] has included a number of technical projects, including: VII Proof-of-Concept (POC) testing, which established the feasibility of the VII system using 5.9 GHz DSRC technology for improving transportation safety and efficiency; Cooperative Intersection Collision Avoidance Systems (CICAS) [8]; and, Vehicle Safety Communications - Applications [9], both of which address automotive collision avoidance applications enabled or enhanced by the deployment of DSRC. Actual testing and equipment used for the research described in this paper were supported in part by the VII proof-of-concept testing activity – a cooperative project by the USDOT and the VII Consortium. The following is the content description of this paper. In the Background Section, we present related work and three typical non-line-of-sight crashes. In the main Section, we describe test settings; vehicle setup, testing locations, and measurement results and discussion, respectively. Finally, we conclude our findings in the Summary Section.

BACKGROUND

Current vehicles are equipped with active safety features such as an anti-lock brake system, vehicle stability assist and track control, and adaptive cruise control with forward looking radar. According to Jiang et al [1], the next logical step in vehicle safety will be incorporating communication based cooperative systems to enhance active safety systems that are already onboard the vehicle. Benefits of communication based safety systems are the ability to warn the driver, and to influence vehicle control systems in limited visibility situations.

Crashes occur for a number of different reasons. Some crashes are due to road and weather conditions especially when these conditions contribute to low visibility. Others occur as a direct, or indirect, result of human errors such as traffic law violation. DSRC technology can be used to reduce crashes by addressing some of the causes, for example: driver inattention, low visibility, and traffic signal violation. By convention, we refer to the offending vehicle as "Host Vehicle" (HV) and other participating vehicles as "Remote Vehicles" (RM). In this paper, we examine communication range and reception rate for three typical crash scenarios: Left Turn Across Path (LTAP), Do Not Pass (DNP) and Intersection Collision Avoidance (ICA).

The Left Turn Across Path crash scenario occurs in the four-way intersection environment. Two vehicles are approaching the intersection from opposite directions. A large truck is in front of the HV. The truck obstructs the line-of-sight of the two vehicles. The truck and HV are turning left. The truck safely completes its left turn. However, the HV, due to lack of visibility, fails to see the RV. A typical LTAP scenario is represented in Figure 1.

Do Not Pass accidents typically happen on the open road. Two vehicles are approaching each other on an open road, driving in opposite directions. The HV is trying to make a passing maneuver around a large truck that is driving in front of the vehicle. The truck obstructs the line-of-sight between the HV and the RV. Due to the truck obstruction, the HV fails to see the RV. Figure 2 shows a typical DNP-type crash.

Common types of Intersection Collision crashes typically occur on four-way intersections. An ICA-type crash is depicted in Figure 3. In this illustration, two vehicles are approaching the intersection. The southwest corner building obstructs line-of-sight.

NON-LINE-OF-SIGHT RELIABILITY MEASUREMENT AND ANALYSIS OF DSRC

EXPERIMENT SETTINGS - We have used two vehicles set up with VII POC hardware and software as shown in Figure 4. Each of the two experimental vehicles is equipped with an On Board Equipment (OBE) with 802.11p-based DSRC radio integrated, an integrated omni-directional roof-mounted DSRC and Global Positioning System (GPS) antenna, a GPS receiver (internal and external to OBE) and a software application in a processor allowing exchange of information via vehicle-to-vehicle communication. Two vehicles are sending the Basic Safety Message (BSM) in wireless short message type, as defined in the Institute of Electrical and Electronics Engineers (IEEE) 802.1p draft standard [10]. Each vehicle sends messages periodically every 100 ms.

The data content of the BSM is defined in Society of Automotive Engineers (SAE) J2735 [11]. The over-the-air packet size includes 45 bytes of overhead, and 39 bytes of data, for a total of 84 bytes.

Messages are sent at 15 dBm with ~4 dB cable loss yielding ~11 dBm. Data rate is 6 Mbps. We have slightly modified the temporary ID field of the BSM to include a sequence number.

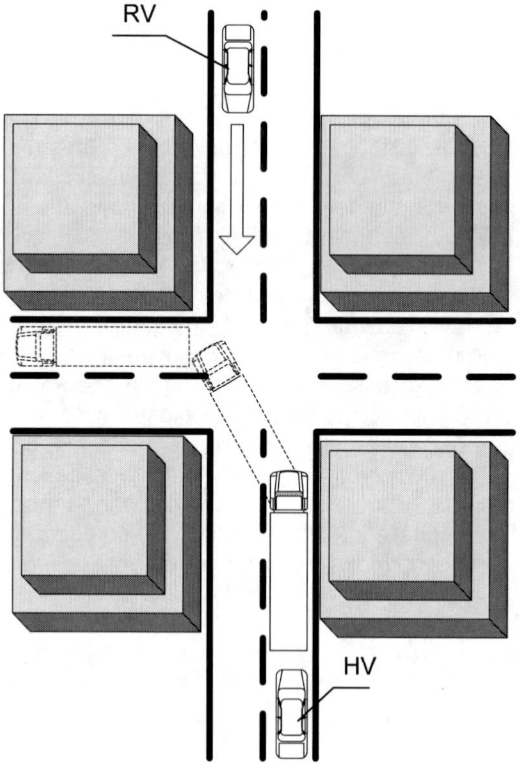

Figure 1 : Visualization of Left Turn Across Path scenario.

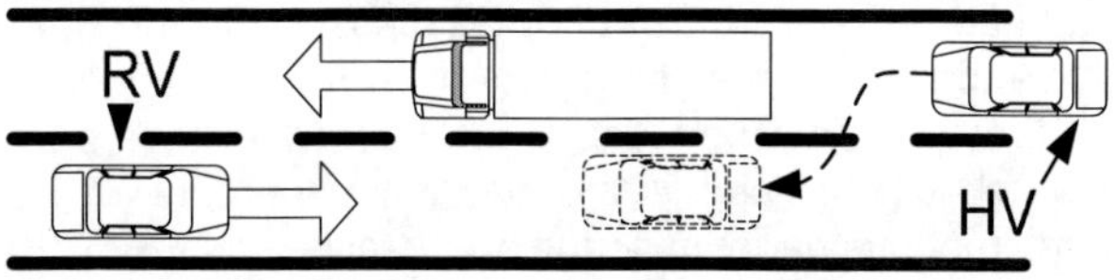

Figure 2 : Visualization of Do Not Pass scenario.

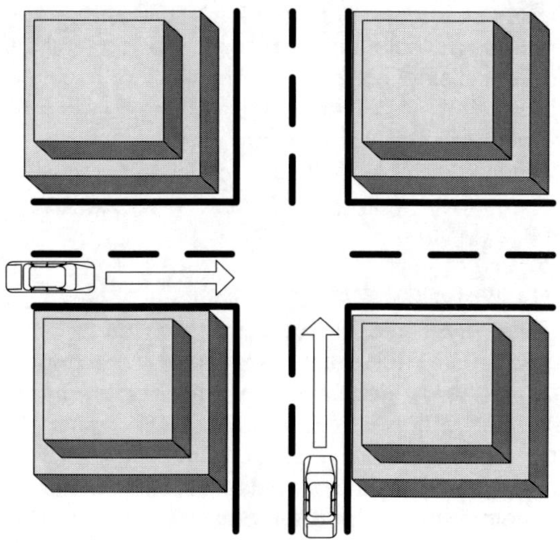

Figure 3 : Visualization of Intersection Collision Avoidance scenario.

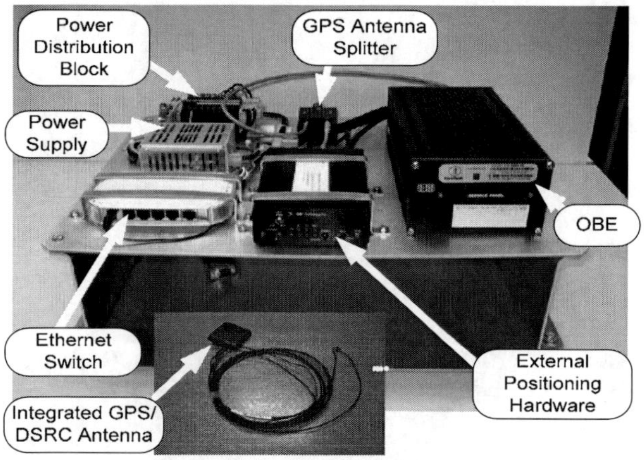

Figure 4 : VII Vehicle's test setup equipment.

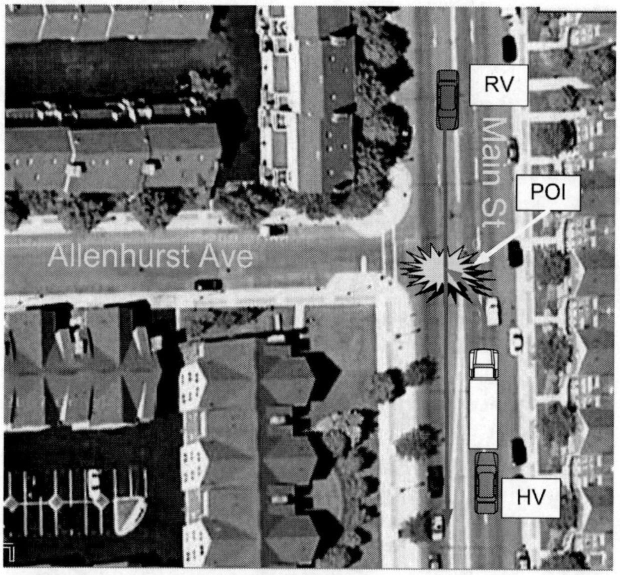

Figure 5 : LTAP scenario location

Figure 6 : DNP scenario location

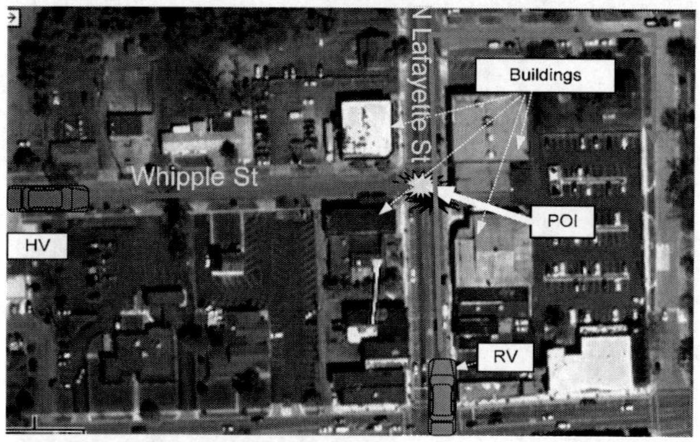

Figure 7 : ICA scenario location

EXPERIMENT LOCATIONS - For investigating the communication range related to the three collision scenarios, we have designed the following test cases. We picked locations that closely resemble the scenarios described previously. The experiments were conducted on April 22 and April 23, 2008 in the Metro Detroit area. The closed intersections provide an environment conducive to DSRC signal multipath.

Figure 5 shows the LTAP scenario location and vehicle positions. The RV is traveling at 35 mph southbound through the intersection. The HV is waiting to turn left on the intersection. The line-of-sight between the HV and the RV is obstructed by a large truck in front of the HV. For our test, we ran 5 LTAP scenario laps.

Figure 6 shows DNP scenario location and vehicle positions. The vehicles are traveling at 55 mph in the opposite direction from each other on the high-speed arterial road. The HV follows a large truck. The large truck in front of the HV obstructs the line-of-sight. We ran ten vehicle passes of the DNP scenario during this part of the testing.

Figure 7 shows the ICA scenario location and vehicle positions. At approximately the same time, the vehicles approach the intersection at 25 mph. The HV sharply stops at the stop sign while the RV goes through the intersection. A corner building obstructs the line-of-sight between the vehicles. For this portion of the testing, we ran 5 ICA scenario laps.

COMMUNICATION RELIABILITY METRICS - Packet Delivery Ratio (PDR) is a widely accepted metric in the literature. PDR is the probability of successfully receiving a packet at the receiver after this packet is transmitted at the sender. PDR is calculated as a ratio of the number of received packets at the receiver to the total number of transmitted packets at the sender within a pre-defined time window. In our experiment, we set this time to be 1 second (i.e. 10 packets). However, the PDR metric communication reliability patterns are purely based on the average value. Here, we also attempt to examine the detailed probability distributions of packet drop patterns across various scenario environments. We believe that this approach might give us a deeper understanding of the impact of potential packet drops on vehicle safety applications. Therefore, we have adopted another statistical metric, Distribution of Consecutive Packet Drops (CPD), as Bai and Krishnan defined in [2], which illustrates the probability distribution of consecutive packet drops for DSRC wireless communication. In addition, we want to examine the range of communication. Therefore, we define Average Distance for First BSM received. We believe this will give us insight into DSRC behavior in non-line-of-sight situations.

EXPERIMENT RESULTS - As described above, we ran multiple laps for each scenario and collected packet logs from the two vehicles. A vehicle log contains transmitted and received messages. Since we had a sequence number embedded in the data content of the message, we used it to find if any messages were missing. The method for extracting reception information from the logs consisted of taking a sequence number from one vehicle transmitted message and finding the corresponding sequence number in the received message in the other vehicle log. If the matching sequence number is found, the corresponding message is considered to have been received; conversely, if the sequence number is not found, the message has not been received. From the logs of the basic safety messages (that includes GPS, time stamp, speed, and sequence number), we computed distance, location, relative speed and received ratio. Received ratio comprises a running average of one second of data. Finally, we averaged out results across multiple laps. We sorted the data samples into different distance bins (the granularity of each bin was set to 4m) and then calculated average packet receive ratio for each distance bin. Finally, we visualized packet received ratio as a function of distance by plotting average received ratio at different distance bins in Figures 8, 9 and 10.

PACKET DELIVERY RATIO – Figure 8 shows LTAP Packet Delivery Ratio vs. distance from the Point-of-Iintersection (POI) for the HV. Since we are primarily interested in the communication behavior before the two vehicles intersect, we separated the PDR curve into incoming and outgoing parts. Incoming PDR is calculated from the data samples collected when the vehicle is approaching the POI. Similarly, outgoing PDR is calculated from the data samples collected when the vehicle is going away from the POI.

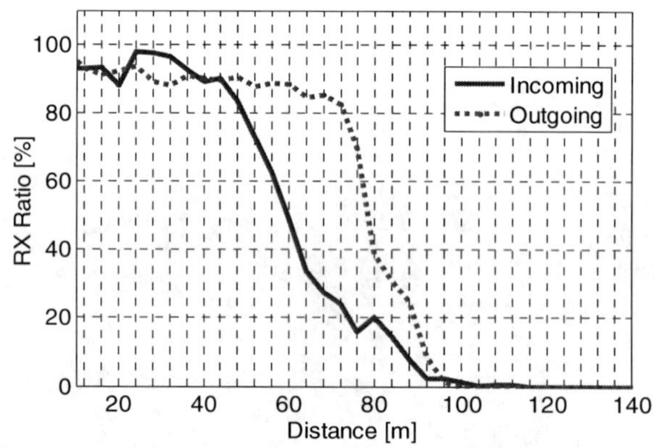

Figure 8 : Host Vehicle: Packet Delivery Ratio vs. Distance Between the Vehicles, for LTAP scenario.

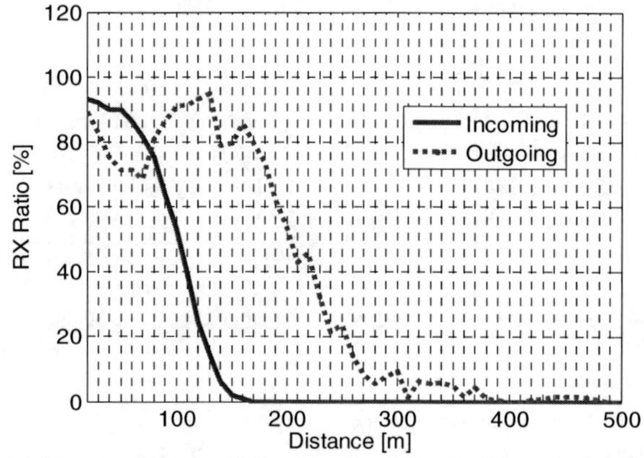

Figure 9 : Host Vehicle: Packet Delivery Ratio vs. Distance between the Vehicles, for DNP scenario

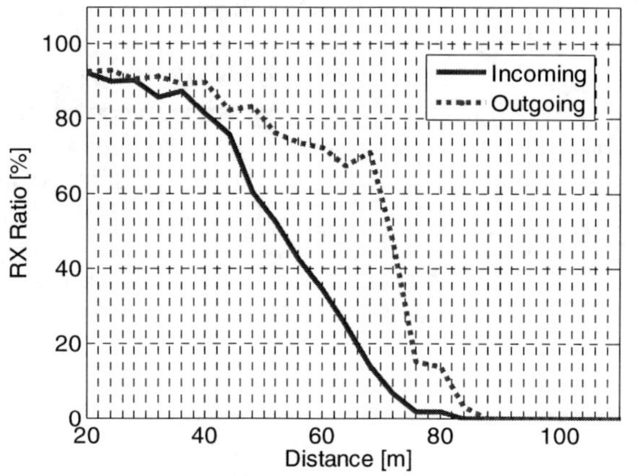

Figure 10 : Host Vehicle: Packet Delivery Ratio vs. Distance Between Vehicles, for ICA scenario

Figure 8 shows that, on average, a PDR (based on packets sent by RV and received by HV) of 70 % or greater is achieved within 52 m from the POI. Assuming that 70 % PDR is sufficient for some safety applications, communication is established, for the static HV and the RV traveling at 25 mph, 4.65 seconds before the RV reaches the POI.

Figure 9 shows PDR versus distance between the vehicles for the DNP scenario. Since we had the vehicles passing each other at slightly different points on the road, we have chosen the POI to be the minimum distance between the vehicles for each run. In this case, a PDR of 70% or greater is reached 81 m from the POI. For the vehicles traveling towards each other at 55 mph, the communication (70 % PDR) is established 1.12 seconds before the vehicles reach the POI.

Figure 10 shows the HV PDR versus distance between vehicles for the ICA scenario. Communication with 70% or greater PDR is reached, on average, for a vehicle distance of 49 m. Assuming that the vehicles travel perpendicularly towards each other at 25 mph, the 70 % PDR mark is reached 3.1 seconds before the vehicles reach the POI.

DISTRIBUTION OF CONSECUTIVE PACKET DROPS - Figure 11 shows Distribution of Consecutive Packet Drops for the three scenarios. The X-axis is the number of consecutive dropped packets and the Y-axis is the probability. This is an important measurement with respect to the envisioned safety applications that might be supported by this system. This importance is based upon the supposition that most of these applications could tolerate a single-packet drop-out, and many could tolerate a two-packet drop-out, without seriously hindering application performance. Three-packet (and higher) drop-outs, however, generally appear to create performance degradation. As shown in Figure 11, most of the packets were dropped one-at-a-time 74%, 69%, and 58% during ICA, LTAP, and DNP testing, respectively , with a lesser percentage being dropped two-at-a-time 22 %, 19 %, and 16% for DNP, ICA, and LTAP testing, respectively. All other burst drops are less than 7 % in all scenarios, which amounts to good news in terms of supporting the envisioned safety applications.

AVERAGE DISTANCE FOR FIRST BSM - The left column of Table I shows the average distance for the first BSM delivered for the three scenarios. The right columns of the table show the percentage of the total packets received for the HV and the RV. The average distance indicates the beginning of communication. Vehicles start communicating well before the line-of-sight, as Figure 12 shows. Red squares indicate no communication, while a green square means the packet is delivered. Non-line-of-sight communication is possible because DSRC waves are reflected by surrounding roads, buildings and trees.

SUMMARY

There is a major interest in assessing the capabilities of DSRC communications to support cooperative collision avoidance safety applications. In order to undertake an experiment within a manageable scope of work, but within radio frequency environments that would be challenging, the communication range and reception rate for three typical crash scenarios: Left Turn Across Path (LTAP), Do Not Pass (DNP) and Intersection Collision Avoidance (ICA) were studied. Prototype VII Proof-of-Concept equipment was used for these studies, with an expectation that production equipment would be likely to provide superior performance over the experimental equipment. The results of this experiment were encouraging. Reliable communications (70% or better packet reception with mainly single-packet drop-outs) were able to be established in the obstructed scenarios selected for LTAP, DNP and ICA applications at distances of 52, 81 and 49 meters. These capabilities, as demonstrated, are expected to be able to support a number of cooperative collision avoidance safety applications that could result in significant safety benefits.

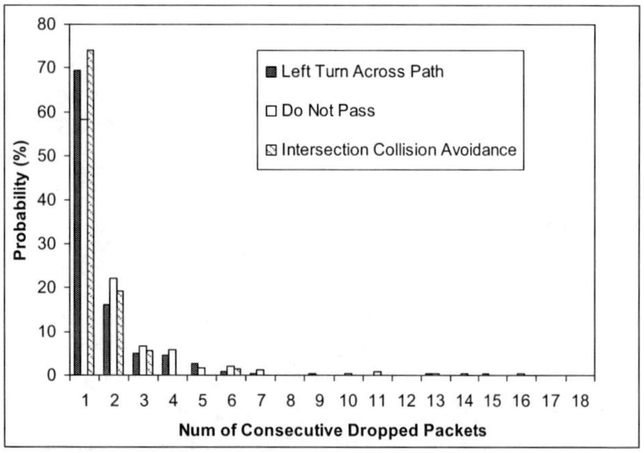

Figure 11 : Probability Distribution Function of Consecutive Packet Drops for LTAP, DNP and ICA

Table I - Average distance for first delivered BSM

	Avg. Distance for 1st BSM [m]	% of BSM Received while in Range	
		HV	RV
LTAP	76.78	77.0	85.5
DNP	97.31	75.6	81.8
ICA	68.11	78.72	87.9

Figure 12 : Average Distance for First BSM for ICA

Additionally, technical improvements to DSRC performance are expected to be possible through development of production-grade equipment and applications software. In conclusion: 1) this prototype DSRC wireless system provided reliable communications, sufficient to support vehicle safety applications, even under the obstructed test scenarios; and 2) future, production grade equipment and applications can be expected to perform more reliably than the prototype test equipment.

ACKNOWLEDGMENTS

The authors would like to thank the VII Consortium and USDOT for providing equipment and resources, as well as the opportunity to participate in the VII proof-of-concept testing. We would also like to thank those VII Consortium technical team members, staff members and contractors who supported these efforts.

REFERENCES

1. Jiang, D.; Taliwal, V.; Meier, A.; Holfelder, W.; Herrtwich, R.; "Design of 5.9 ghz dsrc-based vehicular safety communication," Wireless Communications, IEEE Volume 13, Issue 5, October 2006 Page(s):36 – 43
2. Fan Bai; Krishnan, H.; "Reliability Analysis of DSRC Wireless Communication for Vehicle Safety Applications," Intelligent Transportation Systems Conference, 2006. ITSC '06. IEEE 2006 Page(s):355 – 362
3. Xianbo Chen; Refai, H.H.; Xiaomin Ma; "A Quantitative Approach to Evaluate DSRC Highway Inter-Vehicle Safety Communication," Global Telecommunications Conference, 2007. GLOBECOM '07. IEEE 26-30 Nov. 2007 Page(s):151 – 155
4. Biswas, S.; Tatchikou, R.; Dion, F.; "Vehicle-to-vehicle wireless communication protocols for enhancing highway traffic safety," Communications Magazine, IEEE Volume 44, Issue 1, Jan. 2006 Page(s):74 – 82
5. Mehta, M.; Guinan, M.; "The Utilization of Multi-Antenna Enhanced Mobile Broadband Communications in Intelligent Transportation Systems," Telecommunications, 2007. ITST '07. 7th International Conference on ITS, 6-8 June 2007 Page(s):1 – 4
6. Lott, M.; Meincke, M.; Halfmann, R.; "A new approach to exploit multiple frequencies in DSRC," Vehicular Technology Conference, 2004. VTC 2004-Spring. 2004 IEEE 59th Volume 3, 17-19 May 2004 Page(s):1539 - 1543 Vol.3
7. Vehicle Infrastructure Integration (VII), USDOT Major Initiative, http://www.its.dot.gov/vii/
8. Cooperative Intersection Collision Avoidance Systems (CICAS), USDOT Major Initiative, http://www.its.dot.gov/cicas/index.htm
9. Vehicle Safety Communications Consortium, USDOT Major Initiative, http://www-nrd.nhtsa.dot.gov/pdf/nrd-12/CAMP3/pages/VSCC.htm
10. IEEE, "Wireless Access in Vehicular Environments (WAVE)," Draft 802.1lp/D1.0, New York, NY: IEEE Press, February, 2006.
11. SAE, "Draft SAE j2735 Dedicated Short Range Communications (DSRC) Message Set Dictionary," PA SAE, May 16, 2008

CONTACT

Radovan Miucic and Tom Schaffnit work for Honda R&D Americas, Inc. in the Automobile Technology Research Division. Their research focus is on advanced safety systems enabled by wireless communications. Their address is 1000 Town Center, Suite 2400, Southfield, MI 48075. Radovan can be reached at (248) 304-4701 or via E-mail: rmiucic@oh.hra.com. Tom can be contacted at 248-304-4892, or email tschaffnit@oh.hra.com.

About the Editor

Ronald K. Jurgen

After graduating from Rensselaer Polytechnic Institute with a B.E.E., Ronald K. Jurgen held various technical magazine editorial staff positions, including 30 years with IEEE Spectrum. Now retired, he is the editor of the Automotive Electronics Handbook and the Digital Consumer Electronics Handbook, and assistant editor of the Electronics Engineers' Handbook, Fourth Edition. He is also the editor of more than a dozen SAE International books on automotive electronics.